Astronomers' Universe

Vincent S. Foster

Modern Mysteries of the Moon

What We Still Don't Know About Our Lunar Companion

 Springer

Vincent S. Foster
Waretown, NJ, USA

ISSN 1614-659X   ISSN 2197-6651   (electronic)
Astronomers' Universe
ISBN 978-3-319-22119-9   ISBN 978-3-319-22120-5   (eBook)
DOI 10.1007/978-3-319-22120-5

Library of Congress Control Number: 2015947273

Springer Cham Heidelberg New York Dordrecht London

iStock credit: iStock_000047641846 ©Igor Sokalski

Printed on acid-free paper

Springer International Publishing AG Switzerland is part of Springer Science+Business Media
(www.springer.com)

*Dedicated to the two Kathys in my life,
my wife and my granddaughter*

Preface

Project Apollo was going to be the final chapter of an exciting celestial mystery novel. Scientists were supposed to take the 842 lb of lunar rocks and soil samples brought back from the Moon and with them piece together the origin, evolution, structure, and composition of the Moon. And voila! All the mysteries of the Moon would be solved.

However, reality doesn't always work that way. Data from Apollo and a flotilla of unmanned Moon orbiters, crashers, and landers have yielded many answers, but much remains unknown.

For example, concentric craters are still not fully understood. These odd craters look like someone placed a donut inside the bowl of a crater. What causes them to have such a strange appearance? Take a look at Hesiodus A near Pitatus in Mare Nubium, which is the classic example. Seventy percent of the known concentric craters have morphologies similar to this. Virtually all concentric craters occur on or near maria. The inner slopes of the main crater are smooth and terrace free, similar to normal craters. Only 57 concentric craters are known to exist.

One lunar scientist speculated that their inner rings were volcanic in origin. Another thought that perhaps an effusive flow of relatively viscous magma from fractures at the edge of the crater's floor built up the inner ring. Still another idea was that they are some sort of wall slump feature within normal impact craters. You'll find the surprising answer to this lunar riddle in Chap. 8.

Plato presents another mystery. The gradual darkening of the floor of this crater as the Sun rises has puzzled observers for centuries. Why should the floor of this crater darken while the surrounding areas brighten? Is it due to physical change, as some have speculated?

Another Plato controversy concerns reports that the dark floor is sometimes obscured by mists or clouds. Some of the observations were made during the last century; Walter Goodacre's

1931 book, called *The Moon*, notes that there are "a number of well-authenticated cases." Descriptions include a "curious luminous milky kind of light," "a fog that cleared as the Sun rose," and a nondescript lack of detail. A nineteenth-century observer found that the floor was covered by numerous points of light, "as if reflected from flocculent clouds lying near the surface." These are just some of the mysteries surrounding Plato. They're all discussed in Chap. 4.

The obscuring mist or cloud appearing on Plato may very well be a lunar transient phenomenon (LTP) of a type that has been reported to occur all over the Moon. LTPs are short-lived changes in brightness in areas on the Moon, appearing as quick bright flashes, with some lasting seconds and some lasting for hours! These changes can also involve changing colors such as flashes of red or violet and some have even been recorded as "darkening." These occurrences also have sometimes been described as a haze rather than flash of light, so it appears any occurrence of light on the Moon's surface can be a LTP.

Some LTPs may be caused by gas escaping from underground cavities. These gaseous events are purported to display a distinctive reddish hue, while others have appeared as white clouds or an indistinct haze. The majority of LTPs appears to be associated with floor-fractured craters, the edges of lunar maria, or in other locations linked by geologists with volcanic activity. However, these sites are some of the most popular observing targets, and this correlation could have an observational bias. You'll find other possible answers to these strange phenomena in Chap. 8.

It's also strange that the Moon has been shown to have a "sodium tail" too faint to be detected by the human eye. Hundreds of thousands of miles long, it was discovered in 1998 by Boston University scientists who were observing the Leonid meteor shower. The Moon is constantly releasing atomic sodium gas from its surface, and solar radiation pressure accelerates the sodium atoms in an anti-sunward direction, forming an elongated tail that points away from the Sun. Although it's difficult to see the tail, it's not impossible. Find out how in Chap. 8.

And here's another mystery. Unlike early tests and studies that indicated the Moon had little or no magnetic field, lunar rocks proved to be strongly magnetized. This was shocking to scientists, who could not explain where this magnetic field came from. Later on, scientists were able to determine that when rocks solidify from a lava, they capture a record of the magnetic field in their environment. By studying rocks of different ages, scientists were able to reconstruct the history of lunar surface magnetic fields. This analysis revealed that the intensity of the lunar magnetic field was exceptionally strong 3.56 billion years ago, almost identical to the field measured in the rocks brought back from the Moon. But many questions remain about the Moon's magnetic field, such as why it was so intense late into lunar history and how and why it disappeared over time. The answers are in Chap. 9.

In addition to impact events, the Moon is also rocked by "moonquakes," the lunar equivalent of earthquakes. There are four different types of moonquakes. Deep moonquakes occur up to 700 km below the Moon's surface and are a result of tidal stresses caused by the gravitational tug of war between Earth, the Moon, and the Sun. Shallow moonquakes occur at the surface and down to depths less than about 20 or 30 km, and are often due to landslides of rock down steep crater rims. The Moon also suffers from thermal moonquakes, which occur when the freezing crust expands as it returns to sunlight after 2 weeks of lunar night. Meteorite moonquakes can also cause a rumble or vibration of the surface when a meteoroid slams into the surface. You'll be surprised at which one is the strongest type when you read Chap. 6.

Volcanic activity on the Moon was supposed to have died off centuries ago. That's why astronomers were puzzled when they discovered volcanic activity in the D-shaped volcanic Ina caldera with outgassing and sinkage, leaving relatively bright areas of lunar regolith exposed. The region has to be young in terms of lunar features because of the conspicuous lack of impact craters mottling its surface, like elsewhere on the Moon. You'll find possible answers to this enigma in Chap. 4.

The Aristarchus plateau is one of the most geologically diverse places on the Moon—a mysterious raised flat plateau, a giant rille carved by enormous outpourings of lava, fields of explosive volcanic ash, and all surrounded by a massive flood of basalts. Scientists think the crater was created relatively recently, geologically speaking, when a comet or asteroid smashed into the Moon, gouging out a hole in its surface. The bounty of geologic processes that come together here makes it a high-priority target for future exploration, as explained in Chap. 4.

The "Man in the Moon" was born when cosmic impacts struck the near side of the Moon, the side that faces Earth. These collisions punched holes in the Moon's crust, which later filled with vast lakes of lava that formed the dark areas known as maria, or "seas." In 1959, when the Soviet spacecraft *Luna 3* transmitted the first images of the "dark," or far side, of the Moon, the side facing away from Earth, scientists immediately noticed fewer maria there. This mystery—why no Man in the Moon exists on the Moon's far side—is called the Lunar Farside Highlands Problem. For decades, scientists have been trying to understand why the Earth-facing side of the Moon looks so different from the Moon's far side. Now, they may have an answer to that mystery. See Chap. 10.

The Moon has several bizarre swirl patterns on the surface that look as if visiting aliens took spray paint and left graffiti— "Zord was here." Reiner Gamma is one of the strangest features on the Moon. It is the best-known example of a *lunar swirl* that has baffled planetary geologists for decades. This peculiar pattern looks a *little* like crater rays, but it's curved and lacks a source crater. In fact, there's nothing obviously geological about it. It remains one of the big unanswered questions in lunar science.

The mystery only deepened when orbiting Apollo spacecraft (and later the Lunar Prospector) found strong magnetic fields directly over the swirls—surely an important clue. One theory is that the patchy, localized magnetic field is shunting solar-wind plasma away from the swirls' bright areas and funneling it to the dark lanes that snake through the features. Find out if that's the answer in Chap. 4.

Lunar explorations have revealed that much of the Moon's surface is covered with a glassy glaze, which indicates that the Moon's surface has been scorched by an unknown source of intense heat. As one scientist puts it, the Moon is "paved with glass." One explanation put forward was that an intense solar flare, of awesome proportions, scorched the Moon some 30,000 years or so ago. Scientists have remarked that the glassy glaze is not unlike that created by atomic weapons (the high radioactivity of the Moon should also be considered in light of this theory). Find out more in Chap. 8.

The upper 8 miles of the Moon's crust are surprisingly radio-active. When *Apollo 15* astronauts used thermal equipment, they got unusually high readings, which indicated that the heat flow near the Apennine Mountains was rather hot. The core is not hot at all, but cold. The amount of radioactive materials on the sur-face is not only "embarrassingly high" but difficult to account for. Where did all this hot radioactive material (uranium, thorium, and potassium) come from? Chapter 9 gives you the answer.

Why does the Moon ring like a hollow sphere when a large object hits it? During the Apollo Moon missions, ascent stages of lunar modules as well as the spent third stages of rockets crashed onto the hard surface of the Moon. Each time, these caused the Moon to "ring like a gong or a bell," according to NASA. On the *Apollo 12* flight, reverberations lasted over an hour. Is the Moon hollow? Find out in Chap. 10.

Today, lunar scientists are quick to acknowledge that there is still much to learn about the Moon. In fact, lunar research is an active and healthy field of study. Researchers continually find new ways to dissect the Apollo samples and to interpret old results. A new generation of ever faster supercomputers makes it possible to simulate a wide range of conditions throughout the Moon's lifetime. But even with new ideas, new tools, and new talent, the Moon con-tinues to hold tight to some of its oldest and most cherished secrets.

Waretown, NJ   Vincent S. Foster
July 3, 2014

Acknowledgements

There are many people I would like to thank who have helped to make this book possible. First, I am very grateful to Maury Solomon for her advice and guidance and to her fellow staff members at Springer Publishing who worked so hard to produce this book. I am thankful to Planetary Geologist Chuck Wood for generously sharing his extensive knowledge of the Moon through his *"Lunar Picture of the Day"* website and hundreds of others for their valuable contributions through their books and articles. I am also indebted to the many teams of professional astronomers and planetary geologists whose research is unraveling the many mysteries of the Moon as documented in these pages. Thanks also to NASA and many others for the use of their photographs and illustrations that appear in this book. I am especially grateful to my wife, Kathy, for the patience, understanding, encouragement, and love she gave me without which this book would not have been possible.

Contents

About the Author

Vincent S. Foster has been an amateur astronomer for over 50 years. During that time, he has earned nearly every one of the 40+ observing awards, including the Master's Observer's Award, offered by the Astronomical League. He currently serves as coordinator for two of the League's observing programs—the Hydrogen Alpha Solar Observing Program and the Bright Nebulae Observing Program—both of which he also developed. His favorite celestial object to observe and study is the Moon. He has authored articles for *Astronomy* magazine and the *BBC Sky at Night* magazine. He is a former newspaper reporter. He has been a life-long resident of the New Jersey shore. This is his first book.

1. Views of the Moon from the Past

For most of human history, the Moon has largely been a mystery. Cultures worldwide have long offered up myths to explain its existence. Some ancient peoples believed that the Moon was a rotating bowl of fire. Others thought it was a mirror that reflected Earth's lands and seas. But philosophers in ancient Greece understood that the Moon is a sphere in orbit around Earth. They also knew that moonlight is reflected sunlight.

Some Greek philosophers believed that the Moon was a world much like Earth. In about A. D. 100, Plutarch even suggested that people lived on the Moon. The Greeks also apparently believed that the dark areas of the Moon were seas, while the bright regions were land.

Although Aristotle's natural philosophy was very influential in the Greek world, it was not without competitors and skeptics. Thus, in his little book *On the Face in the Moon's Orb*, the Greek writer Plutarch (46–120 CE) expressed rather different views on the relationship between the Moon and Earth. He suggested that the Moon had deep recesses in which the light of the Sun did not reach and that the spots were nothing but the shadows of rivers or deep chasms. He also entertained the possibility that the Moon was inhabited. In the following century, the Greek satirist Lucian (120–180 CE) wrote of an imaginary trip to the Moon, which was inhabited, as were the Sun and Venus.

In about A. D. 150 Ptolemy, a Greek astronomer who lived in Alexandria, Egypt, said that the Moon was Earth's nearest neighbor in space. He thought that both the Moon and the Sun orbited Earth. Ptolemy's views survived for more than 1300 years.

But by the early 1500s, the Polish astronomer Nicolaus Copernicus had developed the correct view—Earth and the other planets revolve around the Sun, and the Moon orbits Earth.

© Springer International Publishing Switzerland 2016
V.S. Foster, *Modern Mysteries of the Moon*, Astronomers' Universe,
DOI 10.1007/978-3-319-22120-5_1

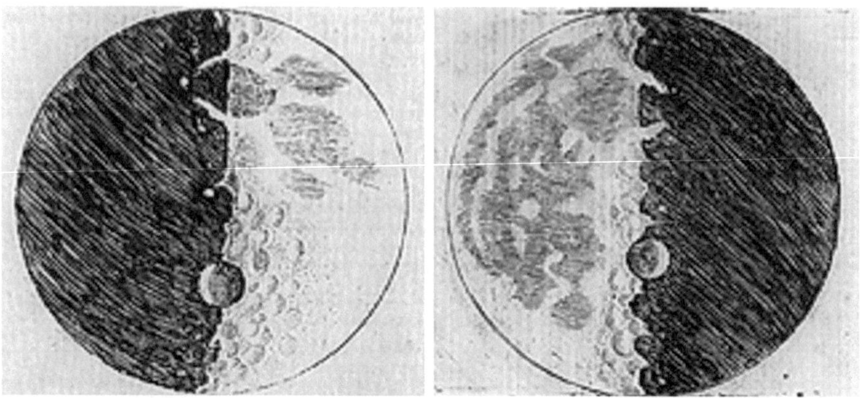

FIG. 1.1 Galileo's sketch of the Moon as he saw it through his telescope. From Galileo's *Sidereus Nuncius* (Starry Messenger). 1610

The Italian astronomer and physicist Galileo wrote the first scientific description of the Moon, based on his observations with a telescope. In 1609, Galileo described a rough, mountainous surface. This description was quite different from what was commonly believed—that the Moon was smooth. Galileo noted that the light regions were rough and hilly and the dark regions were smoother plains. Galileo even sketched the Moon! (Fig. 1.1).

The presence of high mountains on the Moon fascinated Galileo. His detailed description of a large crater in the central highlands—probably Albategnius—began 350 years of controversy and debate about the origin of the "holes" on the Moon.

The medieval followers of Aristotle, first in the Islamic world and then in Christian Europe, tried to make sense of the lunar spots in Aristotelian terms. Various possibilities were entertained. It had already been suggested in antiquity that the Moon was a perfect mirror and that its markings were reflections of earthly features, but this explanation was easily dismissed because the face of the Moon never changes as it moves around Earth. Perhaps there were vapors between the Sun and the Moon, so that the images were actually contained in the Sun's incident light and thus reflected to Earth. The explanation that finally became standard was that there were variations of "density" in the Moon that caused this otherwise perfectly spherical body to appear the way it does. The perfection of the Moon, and therefore the heavens, was thus preserved.

It is a curious fact that although many symbolic images of the Moon survive in medieval and Renaissance works of art (usually a crescent), virtually no one bothered to represent the Moon with its spots the way it actually appeared. We only have a few rough sketches in the notebooks of Leonardo da Vinci (ca. 1500) and a drawing of the naked-eye Moon by the English physician William Gilbert. None of these drawings found its way into print until well after the telescope had come into astronomy (Fig. 1.2).

FIG. 1.2 In 1645, the Dutch engineer and astronomer Michael Florent van Langren, also known as Langrenus, published a map that gave names to the surface features of the Moon, mostly its craters. Of the 325 names that he gave to lunar features, only a handful have been retained (*Source: Plenilunii lumina Austriaca Philippica*, Brussels)

Astronomers of the 1600s mapped and cataloged every surface feature they could see. Increasingly powerful telescopes led to more detailed records. In 1645, the Dutch engineer and astronomer Michael Florent van Langren, published a map that gave names to the surface features of the Moon. A map drawn by the Bohemian-born Italian astronomer Anton M. S. de Rheita in 1645 correctly depicted the bright ray systems of the craters Tycho and Copernicus. Another effort, by the Polish astronomer Johannes Hevelius in 1647, included the Moon's libration zones.

By 1651, two Jesuit scholars from Italy, the astronomer Giovanni Battista Riccioli and the mathematician and physicist Francesco M. Grimaldi, had completed a map of the Moon. That map established the naming system for lunar features that is still in use today.

It was widely believed in the 1800s that the Moon was inhabited. This fantasy was supported by the discovery of what was thought to be a lunar city by the famed Bavarian observer Franz von Paula Gruithuisen. He observed a regular herringbone pattern of ridges arranged with geometric precision in the rough terrain to the north of Schröter Crater near Sinus Aestuum. Gruithuisen immediately interpreted this small landscape (which he called Wallwerk) as artificial structures that he saw as evidence of intelligent life. This interpretation was later dismissed, with the ridges being seen as a natural landform, causing Gruithuisen to lose face (Fig. 1.3).

Until the late 1800s, most astronomers thought that volcanism formed the craters of the Moon. In the 1870s, the English astronomer Richard A. Proctor proposed correctly that the craters resulted from the collision of solid objects with the Moon. But at first, few scientists accepted Proctor's proposal. Most astronomers thought that the Moon's craters must be volcanic in origin because no one had yet described a crater on Earth as an impact crater, but scientists had found dozens of obviously volcanic craters (Fig. 1.4).

In 1892, the American geologist Grove Karl Gilbert argued that most lunar craters were impact craters. He based his arguments on the large size of some of the craters. Those included the basins, which he was the first to recognize as huge craters. Gilbert also noted that lunar craters have only the most general resemblance to calderas (large volcanic craters) on Earth. Both

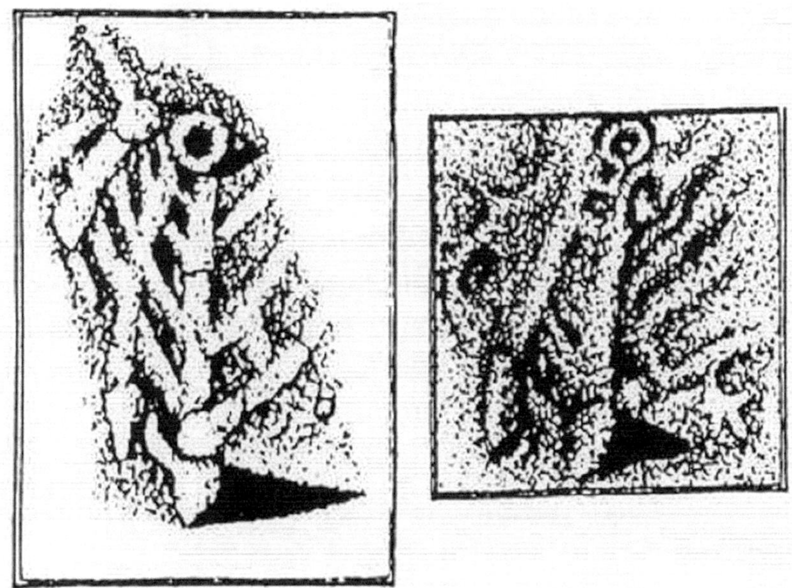

FIG. 1.3 Gruithisen's drawings of his Wallwerk lunar city, including star temple on right, which he regarded as a colossal monument erected by inhabitants of the Moon (From Gruithisen's observations made on July 12, 1822, in Bavaria)

lunar craters and calderas are large circular pits, but their structural details do not resemble each other in any way. In addition, Gilbert created small craters experimentally. He studied what happened when he dropped clay balls and shot bullets into clay and sand targets.

Gilbert was the first to recognize that the circular Mare Imbrium was the site of a gigantic impact. By examining photographs, Gilbert also determined which nearby craters formed before and after that event. For example, a crater that is partially covered by ejecta from the Imbrium impact formed before the impact. A crater within the mare formed after the impact.

Gilbert suggested that scientists could determine the relative age of surface features by studying the ejecta of the Imbrium impact. That suggestion was the key to unraveling the history of the Moon. Gilbert recognized that the Moon is a complex body that was built up by innumerable impacts over a long period.

Fig. 1.4 From the book *Recreations in Astronomy* by H. D. Warren, D. D., published in 1879. The figure was named "Lunar Day," and it represents an historical concept of the lunar surface appearance. Robotic missions to the Moon later demonstrated that the surface features are much more rounded due to a long history of impacts (Illustration courtesy of Henry Warren. Used with permission)

In his book *The Face of the Moon* (1949), the American astronomer and physicist Ralph B. Baldwin further described lunar evolution. He noted the similarity in form between craters on the Moon and bomb craters created during World War II (1939–1945) and concluded that lunar craters form by impact.

Baldwin did not say that every lunar feature originated with an impact. He stated correctly that the maria are solidified flows of basalt lava, similar to flood lava plateaus on Earth. Finally, independently of Gilbert, he concluded that all circular maria are actually huge impact craters that later filled with lava.

In the 1950s, the American chemist Harold C. Urey offered a contrasting view of lunar history. Urey said that, because the

Moon appears to be cold and rigid, it has always been so. He then stated—correctly—that craters are of impact origin. However, he concluded falsely that the maria are blankets of debris scattered by the impacts that created the basins. And he was mistaken in concluding that the Moon never melted to any significant extent. Urey had won the 1934 Nobel Prize in Chemistry and had an outstanding scientific reputation, so many people took his views seriously. Urey strongly favored making the Moon a focus of scientific study. Although some of his ideas were mistaken, his support of Moon study was a major factor in making the Moon an early goal of the U. S. space program.

In 1961, the American geologist Eugene M. Shoemaker founded the branch of astrogeology of the U. S. Geological Survey (USGS). Astrogeology is the study of celestial objects other than Earth. Shoemaker showed that the Moon's surface could be studied from a geological perspective by recognizing a sequence of relative ages of rock layers near the crater Copernicus on the near side. Shoemaker also studied Meteor Crater in Arizona and documented the impact origin of this feature. In preparation for the Apollo missions to the Moon, the USGS began to map the geology of the Moon using telescopes and pictures. This work gave scientists their basic understanding of lunar evolution.

In 1969 the "impossible" came true as the first men walked on the surface of the Moon. For the next 3 years, people of many nationalities watched as one of the great explorations of human history was displayed on their television screens (Fig. 1.5).

Between 1969 and 1972, supported by thousands of scientists and engineers back on Earth, 12 astronauts explored the surface of the Moon. Protected against the airlessness and the killing heat of the lunar environment, they stayed on the Moon for days and some of them traveled for miles across its surface in lunar rovers. They made scientific observations and set up instruments to probe the interior of the Moon. They collected hundreds of pounds of lunar rock and soil, thus beginning the first attempt to decipher the origin and geological history of another world from actual samples of its crust.

The initial excitement of new success and discovery has passed. The TV sets no longer show astronauts moving across the sunlit lunar landscape. But here on Earth, scientists are only now

FIG. 1.5 Shortly after *Apollo 8* orbited the Moon, Pan Am airlines announced plans for commercial flights to the Moon—and they were so confident it would happen soon, they started a waiting list. And so the first Moon flights club was born—attracting more than 93,000 members over the next two decades, each convinced they would soon be following the astronauts into space...just in more comfortable surroundings, with an in-flight magazine and a beverage service, at the very least (Illustration courtesy of Pan Am Historical Foundation)

beginning to understand the immense treasure of new knowledge returned by the Apollo astronauts.

The Apollo program has left us with a large and priceless legacy of lunar materials and data. We now have Moon rocks collected from eight different places on the Moon. The six Apollo landings returned a collection weighing 382 kg (842 lbs.) and consisting of more than 2000 separate samples. Two automated Soviet spacecraft named *Luna-16* and *Luna-20* returned small but important samples totaling about 130 g (5 oz.).

The Apollo program also carried out a major effort of photographing and analyzing the surface of the Moon. Cameras on the Apollo spacecraft obtained so many accurate photographs that we now have better maps of parts of the Moon than we do for some areas on Earth. Special detectors near the cameras measured the weak X-rays and radioactivity given off by the lunar surface. From these measurements, we have been able to determine the chemical composition of about one-quarter of the Moon's surface, an area the size of the United States and Mexico combined. By comparing the flight data with analyses of returned Moon rocks, we can draw conclusions about the chemical composition and nature of the entire Moon.

Although the Apollo program officially ended in 1972, the active study of the Moon goes on. More than 125 teams of scientists are studying the returned lunar samples and analyzing the information that continues to come from the instruments on the Moon. Less than 10 % of the lunar sample material has yet been studied in detail, and more results and discoveries will emerge as new rocks and soil samples are examined. And more questions will arise as well.

Called "the Rosetta Stone of the planets" by Dr. Robert Jastrow, the first chairman of NASA's Lunar Exploration Committee, scientists had hoped that by studying the composition of the Moon it would resolve some of the mysteries of how our planet and Solar System came into existence.

However, six Moon landings later, science writer Earl Ubell declared, "… the lunar Rosetta Stone remains a mystery. The Moon is more complicated than anyone expected; it is not simply a kind of billiard ball frozen in space and time, as many scientists had believed. Some of the fundamental questions have been answered, but the Apollo rocks and recordings have spawned a score of mysteries, a few truly breath-stopping."

2. The Great Moon Hoax

Say the words "Moon hoax" these days, and everyone thinks you are talking about the people who don't believe the Apollo astronauts ever went to the Moon. But back in 1835 there was the original Moon hoax that thousands of people fell for, despite the tall tale being complete fiction. A series of articles were published in the *New York Sun* newspaper reporting incredible new astronomical observations of the Moon supposedly made by astronomer Sir John Herschel during an observing run at the Cape of Good Hope with his powerful new telescope. Detailed descriptions of winged beings, plants, animals and a sapphire temple increased sales and subscriptions to the fledgling newspaper.

Here is a selection of the articles from the series:

GREAT ASTRONOMICAL DISCOVERIES
LATELY MADE
BY SIR JOHN HERSCHEL, L.L.D. F.R.S. & C.
at the Cape of Good Hope
[From Supplement to the Edinburgh Journal of Science]

Day 1

"It was about half past nine o'clock on the night of the tenth, the Moon having then advanced within 4 days of her mean liberation, that the astronomer adjusted his instruments for the inspection of her eastern limb. The whole immense power of his telescope was applied and to its focal image about one half of the power of his microscope. On removing the screen of the latter, the field of view was covered throughout its entire area with a beautifully distinct, and even vivid representation of basaltic rock. Its color was a greenish brown, and the width of the columns, as defined by their interstices on the canvass, was invariably 28 in. No fracture whatever appeared in the mass first presented, but in a few seconds a shelving pile appeared of five or six columns width, which showed

© Springer International Publishing Switzerland 2016
V.S. Foster, *Modern Mysteries of the Moon*, Astronomers' Universe,
DOI 10.1007/978-3-319-22120-5_2

Fig. 2.1 John Herschel's 20-ft telescope at Feldhausen, South Africa, which he reportedly used to discover life on the Moon

their figure to be hexagonal, and their articulations similar to those of the basaltic formation at Staffa. This precipitous shelf was profusely covered with a dark red flower, "precisely similar," says Dr. Grant, "to the Papaver Rhoeas, or rose-poppy of our sublunary cornfields; and this was the first organic production of nature, in a foreign world, ever revealed to the eyes of men." (Fig. 2.1)

"The rapidity of the Moon's ascension, or rather of the Earth's diurnal rotation, being nearly equal to 500 yards in a second, would have effectually prevented the inspection, or even the discovery of objects so minute as these, but for the admirable mechanism which constantly regulates, under the guidance of the sextant, the required altitude of the lens. But its operation was found to be so consummately perfect, that the observers could detain the object upon the field of view for any period they might desire. The specimen of lunar vegetation, however, which they had already seen, had decided a question of too exciting an interest to induce them to retard its exit. It had demonstrated that the Moon has an atmosphere constituted similarly to our own, and capable of sustaining organized, and therefore, most probably animal life. The basaltic rocks continued to pass over the inclined canvass plane, through three successive diameters, when a verdant declivity of great beauty appeared, which occupied two more. This was preceded by another mass of nearly the former height, at the base of which they were at length delighted to perceive that novelty, a lunar forest." "The trees," says Dr. Grant, "for a period of 10 min, were of one unvaried kind, and unlike any I have seen, except the largest kind of yews in the English churchyards, which they in some respects resemble. These were followed by a level green plain, which, as measured by the painted circle on our canvass of 49 ft, must have been more than half a mile in breadth; and then appeared as fine a forest of firs, unequivocal firs, as I have ever seen cherished in the bosom of my native mountains. Wearied with the long continuance of these, we greatly reduced the magnifying power of the microscope, without eclipsing either of the reflectors, and immediately perceived that we had been insensibly descending, as it were, a mountainous district of a highly diversified and romantic character, and that we were on the verge of a lake, or inland sea; but of what relative locality or extent, we were yet too greatly magnified to determine. On introducing the feeblest acromatic lens we possessed, we found that the water, whose boundary we had just discovered, answered in general outline to the Mare Nubium of Riccoli, by which we detected that, instead of commencing, as we supposed, on the eastern longitude of the planet, some delay in the elevation of the great lens had thrown us nearly upon the axis of her equator. However, as she was a free country, and we not, as yet, attached to any particular province, and moreover, since we could at any moment occupy our intended position, we again slid our magic lenses to survey the shores of the Mare Nubium. Why Riccoli so termed it, unless in ridicule of Cleomedes, I know not; for fairer shores never angels coasted on a tour of pleasure." A beach of brilliant white sand, girt with wild castellated rocks, apparently of green marble, varied at chasms, occurring every 200 or 300 ft, with grotesque blocks of chalk or gypsum, and feathered and festooned at the summit with the clustering foliage of unknown trees, moved along the bright wall of our apartment until we were speechless with admiration. The water, we obtained a view of it, was nearly as blue as that of the deep ocean, and broke in large white billows upon the strand. The action of very high tides was quite manifest upon the face of the cliffs for more than a 100 miles; yet diversified as the scenery was during this and a much greater distance, we perceived no trace

FIG. 2.2 Lunarians had glossy copper-colored hair, yellowish skin, and thin membrane wings that extended from their shoulders to their calves. Herschel could tell by their gestures that they were engaged in impassioned speech and were therefore rational and intelligent beings. Herschel's subsequent days of observations found that the man-bats where of different degrees of evolutionary advancement, with some being more evolved than others

of animal existence, notwithstanding we could command at will a perspective or a foreground view of the whole. Mr. Holmes, indeed, pronounced some white objects of a circular form, which we saw at some distance in the interior of a cavern, to be bona fide specimens of a large cornu ammonis; but to me they appeared merely large pebbles, which had been chafed and rolled there by the tides. Our chase of animal life was not yet to be rewarded (Fig. 2.2).

"Having continued this close inspection of nearly 2 h, during which we passed over a wide tract of country, chiefly of a rugged and apparent volcanic character; and having seen few additional varieties of vegetation, except some species of lichen, which grew everywhere in great abundance, Dr. Herschel proposed that we should take out all our lenses, give a rapid speed to the panorama, and search for some of the principal valleys known to astronomers, as the most likely method to reward our first night's observation with the discovery of animated beings. The lenses being removed, and the effulgence of our unutterably glorious reflectors left undiminished, we found in accordance with our calculations, that our field of view comprehended about 25 miles of the lunar surface, with the distinctness both of outline and detail which could be procured of a terrestrial object at a distance of two and a half miles; an optical phenomenon which you will find demonstrated in Note 5. This afforded us the best landscape views we had hitherto obtained, and although the accelerated motion was rather too great, we enjoyed them with rapture. Several of these famous valleys, which are bounded by lofty hills of so perfectly conical a form as to render them less like works of nature than or art, passed the canvass before we had time to check their flight; but presently a train of scenery met our eye, of features so entirely novel, that Dr. Herschel signalled for the lowest convenient gradation of movement. It was a lofty chain of obelisk-shaped, or very slender pyramids, standing in irregular groups, each composed of about thirty or forty spires, every one of which was perfectly square, and as accurately truncated as the finest specimens of Cornish crystal. They were of a faint lilac hue, and very resplendent. I now thought that we had assuredly fallen on productions of art; but Dr. Herschel shrewdly remarked, that if the Lunarians could build 30 or 40 miles of such monuments as these, we should ere now have discovered others of a less equivocal character. He pronounced them quartz formations, of probably wine-colored amethyst species, and promised us, from these and other proofs which he had obtained of the powerful action of laws of crystallization in this planet, a rich field of mineralogical study. On introducing a lens, his conjecture was fully confirmed; they were monstrous amethysts, of a diluted claret color, glowing in the light of the sun! They varied in height from 60 to 90 ft, though we saw several of a still more incredible altitude. They were observed in a succession of valleys divided by longitudinal lines of round-breasted hills, covered with verdure and nobly undulated; but what is most remarkable, the valleys which contained these stupendous crystals were invariably barren, and covered with stones of a ferruginous hue, which were probably iron pyrites. We found that some of these curiosities were situated in a district elevated half a mile above the valley of the Mare Foecunditatis, of Mayer and Riccoli; the shores of which soon hove into view. But never was a name more appropriately bestowed, From "Dan to Bersheba" all was barren, barren—the sea-board was entirely composed of chalk and flint, and not a vestige of vegetation could be discovered with our strongest glasses. The whole breadth of the northern extremity of the sea, which was about 300 miles, having crossed our plane, we entered upon a wild mountainous region abounding with more extensive forests of larger trees than we had seen before—the species of which I have no good analogy to describe. In general contour they resembled our forest oak; but they were much more superb in foliage, having broad glossy leaves like that of the laurel, and tresses of yellow flowers which hung, in the open glades, from the branches to the ground. These mountains passed, we arrived at a region which filled us with utter astonishment. It was an oval valley, surrounded, except at narrow opening towards the south, by hills, red as the purest vermilion, and evidently crystallized; for wherever a precipitous chasm appeared—and these chasms were very frequent, and of immense depth—the perpendicular sections present conglomerated masses of polygon crystals, evenly fitted to each other, and arranged in deep strata, which grew darker in color as they descended to the foundations of the precipices. Innumerable cascades were bursting forth from the breasts of every one of these cliffs, and some so near their summits, and with such great force, as to form arches many yards in diameter. I never was so vividly reminded of Byron's simile, "the tale of the white horse in the Revolution." At the foot of this boundary of hills was a perfect zone of woods surrounding the whole valley, which was about 18 or 20 miles wide, at its greatest breadth, and about 30 in. length. Small collections of trees, of every imaginable kind, were scattered about the whole of the luxuriant area; and here our magnifiers blest our panting hopes with specimens of conscious existence. In the shade of the woods on the south-eastern side, we beheld continuous herds of brown quadrupeds, having all the external characteristics of the bison, but more diminutive than any species of the bos genus in our natural history. Its tail is like that of our bos grunniens; but in its semicircular horns, the hump on its shoulders, and the depth of its dewlap, and the length of its shaggy hair, it closely resembled the species to which I first compared it. It had, however, one widely distinctive feature, which we afterwards found common to nearly every lunar quadruped we have discovered; namely, a remarkable fleshy appendage over the eyes, crossing the whole breadth of the forehead and united to the ears. We could most distinctly perceive this hairy veil, which was shaped like the upper front outline of a cap known to the ladies as Mary Queen of Scots' cap, lifted and lowered by means of the ears. It immediately occurred to the acute mind of Dr. Herschel, that this was a providential contrivance to protect the eyes of the animal from the extremes of light and darkness to which all the inhabitants of our side of the Moon are periodically subjected" (Fig. 2.3).

The next animal perceived would be classed on Earth as a monster. It was of a bluish lead color, about the size of a goat, with a head and beard like him, and a single horn, slightly inclined forward from the perpendicular. The female was destitute of horn and beard, but had a much longer tail. It was gregarious, and chiefly abounded on the acclivitous glades of the woods. In elegance of symmetry it rivalled the antelope, and like him it seemed an agile sprightly creature, running with great speed, and springing from the green turf with all the unaccountable antics of a young lamb or kitten. This beautiful creature afforded us the most exquisite amusement. The mimicry of its movements upon our white painted canvass was as faithful and luminous as that of animals within a few yards of the camera obscrua, when seen pictures upon its tympan. Frequently when attempting to put our fingers upon its beard, it would suddenly bound away into oblivion, as if conscious of our Earthly impertinence; but then others would appear, whom we could not prevent from nibbling the herbage, say or do what we would to them.

On examining the centre of this delightful valley, we found a large branching river, abounding with lovely islands, and waterbirds of numerous kinds. A species of grey pelican was the most numerous; but a black and white crane, with unreasonably long legs and bill, were also quite common. We watched their pisciverous experiments a long time, in hopes of catching sight of a lunar fish; but although we were not gratified in this respect, we could easily guess the purpose with which they plunged their long necks so deeply beneath the water. Near the upper extremity of one of these islands we obtained a glimpse of a strange amphibious creature, of a spherical form, which rolled with great velocity across

FIG. 2.3 A wide variety of extravagant flora and fauna occupied the Moon

the pebbly beach, and was lost sight of in the strong current which set off from this angle of the island. We were compelled, however, to leave this prolific valley unexplored, on account of clouds which were evidently accumulating in the lunar atmosphere, our own being perfectly translucent. But this was itself an interesting discovery, for more distant observers had questioned or denied the existence of any humid atmosphere in this planet.

The Moon being now low on her descent, Dr. Herschel inferred that the increasing refrangibility of her rays would prevent any satisfactory protraction of our labors, and our minds being actually fatigued with the excitement of the high enjoyments we had partaken, we mutually agreed to call in the assistants at the lens, and reward their vigilant attention with congratulatory bumpers of the best "East Indian Particular." It was not, however, without regret that we left the splendid valley of the red mountains, which, in compliment to the arms of our royal patron, we denominated "the Valley of the Unicorn;" and it may be found in Blunt's map, about midway between the Mare Faecunditatis and the Mare Nectaris.

The nights of the eleventh and twelfth being cloudy, were unfavorable to observation; but on those of the thirteenth and fourteenth further animal discoveries were made of the most exciting interest to every human being. We give them in the graphic language of our accomplished correspondent (Fig. 2.4).

FIG. 2.4 Great Moon Hoax lithograph of "ruby amphitheater" for *The Sun*, August 28, 1835 (4th article of 6)

Day 2

"The astonishing and beautiful discoveries which we had made during our first night's observation, and the brilliant promise which they gave of the future, rendered every Moonlight hour too precious to reconcile us to the deprivation occasioned by those two cloudy evenings; and they were borne with strictly philosophical patience, notwithstanding that our attention was closely occupied in superintending the erection of additional props and braces to the 24 ft lens, which we found had somewhat vibrated in a high wind that arose on the morning of the eleventh. The night of the thirteenth (January) was one of pearly purity and loveliness. The Moon ascended the firmament in gorgeous splendor, and the stars, retiring around her, left her the unrivalled queen of the hemisphere. This being the last night but one, in the present month, during which we should have an opportunity of inspecting her western limb, on account of the libration in longitude which would thence immediately ensue, Dr. Herschel informed us that he should direct our resources to the parts numbered 2, 11, 26 and 20 in Blunt's map, and which are respectively known in the modern catalogue by the names of Endymion, Cleomedes, Langrenus, and Petavius. To the careful inspection of these, and the regions between them and the extreme western rim, he proposed to devote the whole of this highly favorable night. Taking then our 25 miles breadth of her surface upon the field of view, and reducing it to a slow movement, we soon found the first very singularly shaped object of our inquiry. It is a highly mountainous district, the loftier chains of which form three narrow ovals, two of which approach each other in slender points, and are united by one mass of hills of great length and elevation; thus presenting a figure similar to that of a long skein of thread, the bows of which have been gradually spread open from their connecting knot. The third oval looks also like a skein, and lies as if carelessly dropped from nature's hand in connection with the other; but that which might fancifully be supposed as having formed the second bow of this second skein is cut open, and lies in scattered threads of smaller hills which cover a great extent of level territory. The ground plan of these mountains is so remarkable that it has been accurately represented in almost every lineal map of the Moon that has been drawn; and in Blunt's, which is the best, it agrees exactly with my description. Within the grasp, as it were, of the broken bow of hills last mentioned, stands an oval-shaped mountain, enclosing a valley of an immense area, and having on its western ridge a volcano in a state of terrific eruption. To the north-east of this, across the broken, or what Mr. Holmes called 'the vagabond mountains,' are three other detached oblong formations, the largest and last of which is marked F in the catalogue, and fancifully denominated the Mare Mortuum, or more commonly the 'Lake of Death.' Induced by a curiosity to divine the reason of so sombre a title, rather than by any more philosophical motive, we here first applied our hydro-oxygen magnifiers to the focal image of the great lens. Our 25 miles portion of this great mountain circus had comprehended the whole of this area, and of course the two conical hills which rise in it about 5 miles from each other; but although this breadth of view had heretofore generally presented its objects as if seen within a terrestrial distance of two and a half miles, we were, in this instance, unable to discern these central hills with any such degree of distinctness. There did not appear to be any mist or smoke around them, as in the case of the volcano which we had left in the south-west, and yet they were completely indistinct upon the canvass. On sliding in the gas-light lens the mystery was immediately solved. They were old craters of extinct volcanoes, from which still issued a heated though transparent exhalation, that kept them in an apparently oscillatory or trembling motion, most unfavorable to examination. The craters of both these hills, as nearly as we could judge under this obstruction, were about 15 fathoms deep, devoid of any appearance of fire, and of nearly a yellowish while color throughout. The diameter of each was about nine diameters of our painted circle, or nearly 450 ft; and the width of the rim surrounding them about 1000 ft; yet notwithstanding their narrow mouths, these two chimneys of the subterranean deep had evidently filled with lava and ashes with which it was encumbered, and even added to the height, if not indeed caused the existence of the oval chain of mountains which surrounded it. These mountains, as subsequently measured from the level of some large lakes around them, averaged the height of 2800 ft; and Dr. Herschel conjectured from this and the vast extent of their abutments, which ran for many miles into the country around them, that these volcanoes must have bee in full activity for a million years. Lieut. Drummond, however, rather supposed that the whole area of this oval valley was but the exhausted crater of one vast volcano, which in expiring had left only these two imbecile representatives of its power. I believe Dr. Herschel himself afterwards adopted this probable theory, which is indeed confirmed by the universal geography of the planet. There is scarcely a 100 miles of her surface, not excepting her largest seas and lakes, in which circular or oval mountainous ridges may not be easily found; and many, very many of these having numerous enclosed hills in full volcanic eruption, which are now much lower than the surrounding circles, it admits of no doubt that each of these great mountains is the remains of one vast mountain which has burnt itself out, and left only these wide formations of its ancient grandeur. A direct proof of this is afforded in a tremendous volcano, now in its prime, which I shall hereafter notice. What gave the name 'The Lake of Death' to the annular mountain I have just described, was, I suppose, the dark appearance of the valley which it encloses, and which, to a more distinct view than we obtained, certainly exhibits the general aspect of the waters on this planet. The surrounding country is fertile to excess: between this circle and No. 2 (Endymion), which we proposed first to examine, we counted not less than 12 luxuriant forests, divided by open plains, which waved in an ocean of vendure, and were probably prairies like those of North America. In three of these we discovered numerous herds of quadrupeds similar to our friends the bisons in the Valley of the Unicorn, but of much larger size; and scarcely a piece of woodland occurred in our panorama which did not dazzle our visions with flocks of white or red birds upon the wing (Fig. 2.5).

"At length we carefully explored the Endymion. We found each of the three ovals volcanic and sterile within; but, without, most rich, throughout the level regions around them, in every imaginable production of a bounteous soil. Dr. Herschel has classified not less than 38 species of forest trees, and nearly twice this number of plants, found in this tract alone, which are widely different to those found in more equatorial latitudes. Of animals, he classified nine species of mammalia, and five of ovipara. Among the former is a small kind of rein-deer, the elk, the moose, the horned bear, and the biped beaver. The last resembles the beaver of the Earth in every other respect than in its destitution of a tail, and its invariable habit of walking upon only 2 ft. It carries its young

FIG. 2.5 In the midst of a valley near a beautiful lake, there was a wide expanse of plants where Herschel spotted four flocks of large winged creatures. It was at this time that he proclaimed to his assistant, Dr. Andrew Grant, that here is where one would expect to find signs of advanced life and with this proclamation he did indeed discover lunar sentient life. Upon changing to a different lens, Herschel had his first close-up encounter with what appeared to be bat-winged humanoid creatures walking upright, hence given the Latin name *Vespertilio-homo*

in its arms like a human being, and moves with an easy gliding motion. Its huts are constructed better and higher than those of many tribes of human savages, and from the appearance of smoke in nearly all of them, there is no doubt of its being acquainted with the use of fire. Still its head and body differ only in the points stated from that of the beaver, and it was never seen except on the borders of lakes and rivers, in which is has been seen to immerse for a period of several seconds".

"Thirty degrees farther south, in No. 11, or Cleomedes, an immense annular mountain, containing three distinct craters, which have been so long extinguished that the whole valley around them, which is 11 miles in extent, is densely crowded with woods nearly to the summits of the hills. Not a rod of vacant land, except the tops of these craters, could be descried, and no living creature, except a large white bird resembling the stork. At the southern extremity of this valley is a natural archway or cavern, 200 ft high, and 100 wide, through which runs a river which discharges itself over a precipice of grey rock 80 ft in depth, and thus forms a branching stream through a beautiful campaign district for many miles. Within 12 miles of this cataract is the largest lake, or rather inland sea, that has been found throughout the seven and a half millions of square miles which this illuminated side of the Moon contains. Its width, from east to west, is 198 miles, and from north to south, 266 miles. Its shape, to the northward, is not unlike that of the Bay of Bengal, and it is studded with small islands, most of which are volcanic. Two of these, on the eastern side, are now violently eruptive; but our lowest magnifying power was too great to examine them with convenience, on account of the cloud of smoke and ashes which beclouded our field of view: as seen by Lieut. Drummond, through our reflective telescope of 2000 times, they exhibited great brilliancy. In a bay, on the western side of this sea, is an island 55 miles long, of a crescent form, crowded through its natural sweep with the most superb and wonderful natural beauties, both of vegetation and geology. Its hills are pinnacled with tall quartz crystals, of so rich a yellow and orange hue that we at first supposed them to be pointed flames of fire; and they spring up thus from smooth round brows of hills which are covered with a velvet mantle. Even in the enchanting little valleys of this winding island we could often see these splendid natural spires, mounting in the midst of deep green woods, like church steeples in the vales of Westmoreland. We here first noticed the lunar palm-tree, which differs from that of our tropical latitudes only in the peculiarity of very large crimson flowers, instead of the spadix protruded from the common calyx. We, however, perceived no fruit on any specimens we saw: a circumstance which we attempted to account for from the great (theoretical) extremes in the lunar climate. On a curious kind of tree-melon we nevertheless saw fruit in great abundance, and in every stage of inception and maturity. The general color of these woods was a dark green, though not without occasional admixtures of every tint of our forest seasons. The hectic flush of autumn was often seen kindled upon the cheek of earliest spring; and the gay drapery of summer in some places surrounded trees leafless as the victims of winter. It seemed as if all the seasons here united hands in a circle of perpetual harmony. Of animals we saw only an elegant striped quadruped about 3 ft high, like a miniature zebra; which was always in small herds on the green sward of the hills; and two or three kinds of long-tailed birds, which we judges to be golden and blue pheasants. On the shores, however, we saw countless multitudes of univalve shell-fish, and among them some huge flat ones, which all three of my associates declared to be cornu ammonae; and I confess I was here compelled to abandon my sceptical substitution of pebbles. The cliffs all along these shores were deeply undermined by tides; they were very cavernous, and yellow crystal stalactites larger than a man's thigh were shooting forth on all sides. Indeed every rood of this island appeared to be crystallized; masses of fallen crystals were found on every beach we explored, and beamed from every fractured headland. It was more like a creation of an oriental fancy than a distinct variety of nature brought by the powers of science of ocular demonstration. The striking dissimilitude of this island to every other we had found on these waters, and its near proximity to the main land, led us to suppose that it must at some time have been a part of it; more especially as its crescent bay embraced the first of a chain of smaller ones which ran directly thither. The first one was a pure quartz rock, about 3 miles in circumference, towering in naked majesty from the blue deep, without either shore or shelter. But it glowed in the sun almost like a sapphire, as did all the lesser ones of whom it seemed the king. Our theory was speedily confirmed; for all the shore of the main land was battlemented and spired with these unobtainable jewels of nature; and as we brought our field of view to include the utmost rim of the illuminated boundary of the planet, we could still see them blazing in crowded battalions as it were, through a region of 100 of miles. If fact we could not conjecture where this gorgeous land of enchantment terminated; for as the rotary motion of the planet bore these mountain summits from our view, we became further remote from their western boundary."

"We were admonished by this to lose no time in seeking the next proposed object of our search, the Langrenus, or No. 26, which is almost within the verge of the libration in longitude, and of which, for this reason, Dr. Herschel entertained some singular expectations" (Fig. 2.6).

Fig. 2.6 After a time of observation, Dr. Herschel passed the Valley of the Unicorn where he found more of the horned bison-like creatures from the previous day's observation. In addition to spotting the bison, Herschel found several other mammals including: a small type of reindeer, elk, a horned bear, moose and a biped beaver. The most remarkable of all discoveries was the bipedal beaver. These remarkable creatures lived in huts, had knowledge of making fires, and carried their young much like humans do. The presence of these beavers, which walked on two legs and had knowledge of basic technology, gave hope that even more advanced creatures could be found, maybe even ones sentient and humanoid in nature

Day 3

"After a short delay in advancing the observatory upon the levers, and in regulating the lens, we found our object and surveyed it. It was a dark narrow lake 70 miles long, bounded, on the east, north, and west, by red mountains of the same character as those surrounding the Valley of the Unicorn, from which it is distant to the south-west about 160 miles. This lake, like that valley, opens to the south upon a plain not more than 10 miles wide, which is here encircled by a truly magnificent amphitheater of the loftiest order of lunar hills. For a semicircle of 6 miles these hills are riven, from their brow to their base, as perpendicularly as the outer walls of the Colosseum at Rome; but here exhibiting the sublime altitude of at least 2000 ft, in one smooth unbroken surface. How nature disposed of the large mass which she thus prodigally carved out, I know not; but certain it is that there are no fragments of it left upon the plain, which is a declivity without a single prominence except a billowly tract of woodland that runs in many a will vagary of breadth and course to the margin of the lake. The tremendous height and expansion of this perpendicular mountain, with its bright crimson front contrasted with the fringe of forest on its brow, and the verdure of the open plain beneath, filled out canvass with a landscape unsurpassed in unique grandeur by any we had beheld. Our 25 miles perspective included this remarkable mountain, the plain, a part of the lake, and the last graduated summits of the range of hills by which the latter is nearly surrounded. We ardently wished that all the world could view a scene so strangely grand, and our pulse beat high with the hope of one day exhibiting it to our countrymen in some part of our native land. But we were at length compelled to destroy our picture, as a while, for the purpose of magnifying its parts for scientific inspection. Our plain was of course immediately covered with the ruby front of this mighty amphitheater, its tall figures, leaping cascades, and rugged caverns. As its almost interminable sweep was measured off on the canvass, we frequently saw long lines of some yellow metal hanging from the crevices of the horizontal strata in will net-work, or straight pendant branches. We of course concluded that this was virgin gold, and we had no assay-master to prove to the contrary. On searching the plain, over which we had observed the woods roving in all the shapes of clouds in the sky, we were again delighted with the discovery of animals. The first observed was a quadruped with an amazingly long neck, head like a sheep, bearing two long spiral horns, white as polished ivory, and standing in a perpendicular parallel to each other. Its body was like that of a deer, but its fore-legs were most disproportionally long, and its tail, which was very busy and of a snowy whiteness, curled high over its rump, and hung 2 or 3 ft by its side. Its colors were bright bay and white in brindled parches, clearly defined, but of no regular form. It was found only in pairs, in spaces between the woods, and we had no opportunity of witnessing its speed or habits. But a few minutes only elapsed before three specimens of another animal appeared, so well known to us all that we fairly laughed at the recognition of so familiar an acquaintance in so distant a land. They were neither more nor less than three good large sheep, which would not have disgraced the farms of Leicestershire, or the shambles of Leanenhall-market. With the utmost scrutiny, we could find no mark of distinction between these and those of our native soil; they had not even the appendage over the eyes, which I have described as common to lunar quadrupeds. Presently they appeared in great numbers, and on reducing the lenses, we found them in flocks over a great part of the valley. I need not say how desirous we were of finding shepherds to these flocks, and even a man with blue apron and rolled up sleeves would have been a welcome sight to us, if not to the sheep; but they fed in peace, lords of their own pastures, without either protector or destroyer in human shape" (Fig. 2.7).

"We at length approached the level opening to the lake, where the valley narrows to a mile in width, and displays scenery on both sides picturesque and romantic beyond the powers of prose description. Imagination, borne on the wings of poetry, could alone gather similes to portray the wild sublimity of this landscape, where dark behometh crags stood over the brows of lofty precipices, as if a rampart in the sky; and forests seemed suspended in mid air. On the eastern side there was one soaring crag, crested with trees, which hung over in a curve like three-fourths of a Gothic arch, and being of a rich crimson color, its effect was most strange upon minds unaccustomed to the association of such grandeur with such beauty. "But whilst gazing upon them in a per-

FIG. 2.7 Along with iconic images of the man-bats, horned bison, and bipedal beavers, an Italian version of the newspaper story contains nineteenth century missionaries on the Moon studying and capturing the Lunarians and a beautiful balloon used by them to travel to the Moon. These additions have interesting back stories. The inclusion of the explorers and missionaries could have been the result of a general public call to send missionaries and bibles to the Moon to make sure the Lunarians were well versed in the Bible

spective of about half a mile, we were thrilled with astonishment to perceive four successive flocks of large winged creatures, wholly unlike any kind of birds, descend with a slow even motion from the cliffs on the western side, and alight upon the plain. They were first noted by Dr. Herschel, who exclaimed, "Now, gentlemen, my theories against your proofs, which you have often found a pretty even bet, we have here something worth looking at: I was confident that if we ever found beings in human shape, it would be in this longitude, and that they would be provided by their Creator with some extraordinary powers of locomotion: first exchange for my number D.' This lense being soon introduced, gave us a fine half-mile distance, and we counted 3 parties of these creatures, of 12, none, and 15 in each, walking erect towards a small wood near the base of the eastern precipices. Certainly the were like human beings, for their wings had now disappeared, and their attitude in walking was both erect and dignified. Having observed them at this distance for some minutes, we introduced lens Hz which brought them to the apparent proximity of 80 yards; the highest clear magnitude we possessed until the latter end of March, when we effected an improvement in the gas-burners. About half of the first party had passed beyond our canvass; but of all the others we had a perfect distinct and deliberate view. They averaged 4 ft in height, were covered, except on the face, with short and glossy copper-colored hair, and had wings composed of a thin membrane, without hair, lying snugly upon their backs, from the top of their shoulders to the calves of their legs. The face, which was of a yellowish flesh color, was a slight improvement upon that of the large orang outang, being more open and intelligent in its expression, and having a much greater expansion of forehead. The mouth, however, was very prominent, though somewhat relieved by a thick beard upon the lower jaw, and by lips far more human than those of any species of simia genus. In general symmetry of body and limbs they were infinitely superior to the orang outang; so much so, that, but for their long wings, Lieut. Drummond said they would look as well on a parade ground as some of the old cockney militia! The hair on the head was a darker color than that of the body, closely curled, but apparently not wooly, and arranged in two curious semicircles over the temples of the forehead. Their feet could only be seen as they were alternately lifted in walking;

but, from what we could see of them in so transient a view, they appeared thin, and very protuberant at the heel."

"Whilst passing across the canvas, and whenever we afterwards saw them, these creatures were evidently engaged in conversation; their gesticulation, more particularly the varied action of their hands and arms, appeared impassioned and emphatic. We hence inferred that they were rational beings, and although not perhaps of so high an order as others which we discovered the next month on the shores of the Bay of Rainbows, they were capable of producing works of art and contrivance. The next view we obtained of them was still more favorable. It was on the borders of a little lake, or expanded stream, which we then for the first time perceived running down the valley to a large lake, and having on its eastern margin a small wood. "Some of these creatures had crossed this water and were lying like spread eagles on the skirts of the wood. We could then perceive that they possessed wings of great expansion, and were similar in structure to this of the bat, being a semi-transparent membrane expanded in curvilineal divisions by means of straight radii, united at the back by the dorsal integuments. But what astonished us very much was the circumstance of this membrane being continued, from the shoulders to the legs, united all the way down, though gradually decreasing in width. The wings seemed completely under the command of volition, for those of the creatures whom we saw bathing in the water, spread them instantly to their full width, waved them as ducks do their to shake off the water, and then as instantly closed them again in a compact form. Our further observation of the habits of these creatures, who were of both sexes, led to results to very remarkable, that I prefer they should first be laid before the public in Dr. Herschel's own work, where I have reason to know they are fully and faithfully stated, however incredulously they may be received.—The three families then almost simultaneously spread their wings, and were lost in the dark confines of the canvass before we had time to breathe from our paralyzing astonishment. We scientifically denominated them as Vespertilio-homo, or man-bat; and they are doubtless innocent and happy creatures, notwithstanding that some of their amusements would but ill comport with our terrestrial notions of decorum. The valley itself we called the Ruby Coloseum, in compliment to its stupendous southern

FIG. 2.8 Pictures showing daily activity of the Lunarians

boundary, the 6 mile sweep of precipices 2000 ft high. And the night, or rather morning, being far advanced, we postponed our tour to Petavius (No. 20), until another opportunity." (Fig. 2.8)

"We have, of course, faithfully obeyed Dr. Grant's private injunction to omit those highly curious passages in his correspondence which he wished us to suppress, although we do not perceive the force of the reason assigned for it. It is true, the omitted paragraphs contain facts which would be wholly incredible to readers who do not carefully examine the principles and capacity of the instrument with which these marvellous discoveries have been made; but so will nearly all those which he has kindly permitted us to publish; and it was for this reason we considered the explicit description which we have given of the telescope so important a preliminary. From these, however, and other prohibited passages, which will be published by Dr. Herschel, with the certificates of the civil and military authorities of the colony, and of several Episcopal, Wesleyan, and other ministers, who, in the month of March last, were permitted, under the stipulation of temporary secrecy, to visit the laboratory, and become eye-witnesses of the wonders which they were requested to attest, we are confident his forthcoming volumes will be at once the most sublime in science, and the most intense in general interest, that ever issued from the press.

The night of the fourteenth displayed the Moon in her mean libration, or full; but the somewhat humid state of the atmosphere being for several hours less favorable to a minute inspection than to a general survey of her surface, they were chiefly devoted to the latter purpose. But shortly after midnight the last veil of mist was dissipated, and the sky being as lucid as on the former evenings, the attention of the astronomers was arrested by the remarkable outlines of the spot marked Tycho, No. 18, in Blunt's lunar chart; and in this region they added treasures to human knowledge which angels might well desire to win. Many parts of the following extract will remain forever in the chronicles of time."

Day 4

"The surface of the Moon, when viewed in her mean libration, even with telescopes of very limited power, exhibits three oceans of vast breadth and circumference, independently of seven large collections of water, which may be denominated seas. Of inferior waters, discoverable by the higher classes of instruments, and usually called lakes, the number is so great that no attempt has yet been made to count them. Indeed, such a task would be almost equal to that of ennumerating the annular mountains which are found upon every part of her surface, whether composed of land or water. The largest of the three oceans occupies a considerable portion of the hemisphere between the line of her northern axis and that of her eastern equator, and even extends many degrees south of the latter. Throughout its eastern boundary, it so closely approaches that of the lunar sphere, as to leave in many places merely a fringe of illuminated mountains, which are here, therefore, strongly contra-distinguished from the dark and shadowy aspect of the great deep. But peninsulas, promontories, capes, and islands, and a thousand other terrestrial figures, for which we can find no names in the poverty of our geographical nomenclature, are found expanding, sallying forth, or glowing in insular independence, through all the 'billowy boundlessness' of this magnificent ocean. One of the most remarkable of these is a promontory, without a name, I believe, in the lunar charts, which starts from an island district denominated Copernicus by the old astronomers, and abounding, as we eventually discovered, with great natural curiosities. This promontory is indeed most singular. Its northern extremity is shaped much like an imperial crown, having a swelling bow, divided and tied down in its centre by a band of hills which is united with its forehead or base. The two open spaces formed by this division are two lakes, each 80 miles wide; and at the foot of these, divided from them by the band of hills last mentioned, is another lake, larger than the two put together, and nearly perfectly square. This one is followed, after another hilly division, by a lake of an irregular form; and this one yet again, by two narrow ones, divided longitudinally, which are attenuated northward to the main land. Thus the skeleton promontory of mountain ridges runs 396 miles into the ocean, with six capacious lakes, enclosed within its stony ribs. Blunt's excellent lunar chart gives this great work of nature with wonderful fidelity, and I think you might accompany my description with an engraving from it, much to your reader's satisfaction."

"Next to this, the most remarkable formation in this ocean is a strikingly brilliant annular mountain of immense altitude and circumference, standing 330 miles E.S.E., commonly known as Aristarchus (No. 12), and marked in the chart as a large mountain, with a great cavity in its centre. That cavity is, now, as it was probably wont to be in ancient ages, a volcanic crater, awfully rivaling our Mounts Etna and Versuvius in the most terrible epochs of their reign. Unfavorable as the state of the atmosphere was to close examination, we could easily mark its illumination of the water over a circuit of 60 miles. If we have before retained any doubt of the power of lunar volcanoes to throw fragments of their craters so far beyond the Moon's attraction that they would necessarily gravitate to this Earth, and thus account for the multitude of massive aerolites which have fallen and been found upon our surface, the view which we had of Aristarchus would have set our skepticism forever at rest. This mountain, however, though standing 300 miles in the ocean, is not absolutely insular, for it is connected with the main land by four chains of mountains, which branch from it as a common centre."

"The next great ocean is situated on the western side of the meridian line, divided nearly in the midst by the line of the equator, and is about 900 miles in north and south extent. It is marked C in the catalogue, and was fancifully called the Mare Tranquillitatis. It is rather two large seas than one ocean, for it is narrowed just under the equator by a strait not more than 100 miles wide. Only three annular islands of a large size, and quite detached from its shores, are to be found within it; though several sublime volcanoes exist on its northern boundary; one of the most stupendous of which is within 120 miles of the Mare Nectaris before mentioned. Immediately contiguous to this second great ocean, and separated from it only by a concatenation of dislocated continents and islands, is the third, marked D, and known as the Mare Serenitatis. It is nearly square, being about 330 miles in length and width. But is has one most extraordinary peculiarity, which is a perfectly straight ridge of hills, certainly not more than 5 miles wide, which starts in a direct line from its southern to its northern shore, dividing it exactly in the midst. This singular ridge is perfectly sui generis, being altogether unlike any mountain chain either on this Earth or on the Moon itself. It is so very keen, that its great concentration of the solar light renders it visible to small telescopes; but its character is so strikingly peculiar, that we could not resist the temptation to depart from our predetermined adherence to a general survey, and examine it particularly. Our lens Gx brought it within the small distance of 800 yards, and its whole width of 4 or 5 miles snugly within that of our canvass. Nothing that we had hitherto seen more highly excited our astonishment. Believe it or believe it not, it was one entire crystallization!—its edge, throughout its whole length of 340 miles, is an acute angle of solid quartz crystal, brilliant as a piece of Derbyshore spar just brought from a mine, and containing scarcely a fracture or a chasm from end to end! What a prodigious influence must our 13 times larger globe have exercised upon this satellite, when an embryo in the womb of time, the passive object of chemical affinity! We found that wonder and astonishment, as excited by objects in this distant world, were but modes and attributes of ignorance, which should give place to elevated expectations, and to reverential confidence in the illimitable power of the Creator" (Fig. 2.9).

"The dark expanse of waters south of the first great ocean has often been considered a fourth; but we found it to be merely a sea of the first class, entirely surrounded by land, and much more encumbered with the promontories and islands that it has been exhibited in any lunar chart. One of its promontories runs from the vicinity of Pitatus (No. 19), in a slightly curved and very narrow line, to Bullialdus (No. 22), which is merely a circular head to it, 264 miles from its starting place. This is another mountainous ring, a marine volcano, nearly burnt out, and slumbering upon its cindres. But Pictatus, standing upon a bold cape of the southern shore, is apparently exulting in the might and majesty of its fires. The atmosphere being now quite free from vapor, we introduced the magnifiers to examine a large bright circle of hills which sweep close beside the western abutments of this flaming mountain. The hills were either of snow-white marble or semi-transparent crystal, we could not distinguish which, and they bounded another of those lovely green valleys, which, however monotonous in my descriptions, are of paradisiacal beauty and fertility, and like primitive Eden in the bliss of their inhabitants. Dr. Herschel again predicted another of his sagacious theories. He said the proximity of the flaming mountain, Bullialdus, must be so great a local convenience to dwellers in this valley during the long periodical absence of solar light, as to render it a place of populous resort for inhabitants of all the adjacent regions, more especially as its bulwark of hills afforded an infallible security against any volcanic eruptions that could occur. We therefore applied our full power to explore it, and rich indeed was our reward."

FIG. 2.9 Portrait of a man-bat (*"Vespertilio-homo"*), from an edition of
the Moon series published in the *New York Sun*

"The very first object in this valley that appeared upon our
canvass was a magnificent work of art. It was a temple—a fane of
devotion, or of science, which, when consecrated to the Creator is
devotion of the loftiest order; for it exhibits his attributes purely
free from the masquerade, attire, and blasphemous caricature of
controversial creeds, and has the seal and signature of his own
hand to sanction its aspirations. It was an equitriangular temple,
built of polished sapphire, or of some resplendent blue stone,
which, like it, displayed a myriad points of golden light twinkling

and scintillating in the sunbeams. Our canvass, though 50 ft in diameter, was too limited to receive more than a sixth part of it at one view, and the first part that appeared was near the centre of one of its sides, being three square columns, 6 ft in diameter at its base, and gently tapering to a height of 70 ft. The intercolumniations were each 12 ft. We instantly reduced our magnitude, so as to embrace the whole structure in one view, and then indeed it was most beautiful. The roof was composed of some yellow metal, and divided into three compartments, which were not triangular planes inclining to the centre, but subdivided, curbed, and separated, so as to present a mass of violently agitated flames rising from a common source of conflagration and terminating in wildly waving points. This design was too manifest, and too skillfully executed to be mistaken for a single monument. Though a few openings in these metallic flames we perceived a large sphere of a darker kind of metal nearly of a clouded copper color, which they enclosed and seemingly raged around, as if hieroglyphically consuming it. The was the roof; but upon each of the three corners there was a small sphere of apparently the same metal as the large centre one, and these rested upon a kind of cornice, quite new in any order of architecture with which we are acquainted, but nevertheless exceedingly graceful and impressive. It was a half-opened scroll, swelling off boldly from the roof, and hanging far over the walls in several convolutions. It was of the same metal as the flames, and on each side of the building it was open at both ends. The columns, six on each side, were simply plain shafts, without capitals or pedestals, or any description of ornament; nor was any perceived in other parts of the edifice. It was open on each side, and seemed to contain neither seats, altars, nor offerings; but it was a light and airy structure, nearly a 100 ft high from its white glistening floor to its glowing roof, and it stood upon a round green eminence on the eastern side of the valley. We afterwards, however, discovered two others, which were in every respect facsimiles of this one; but in neither did we perceive and visitants besides flocks of wild doves which alighted upon its lustrous pinnacles. Had the devotees of these temples gone the way of all living, or were the latter merely historical monuments? What did the ingenious builders mean by the globe surrounded by flames? Did they by this record any past calamity of their world, or predict any future one of ours? I by no means despair of ultimately solving not only these but a thousand other questions which present themselves respecting the objects of this planet; for not the millionth part of her surface has yet been explored, and we have been more desirous of collecting the greatest possible number of new facts, than of indulging in speculative theories, however seductive for the imagination."

Day 5

"But we had far not to seek for inhabitants of this 'Vale of the Triads.' Immediately on the outer border of the wood which surrounded, at a distance of half a mile, the eminence on which the first of these temples stood, we saw several detached assemblies of beings whom we instantly recognized to be of the same species as our winged friends of the Ruby Colosseum near the lake Langrenus. Having adjusted the instrument for a minute examination, we found that nearly all the individuals in these groups were of larger stature than the former specimens, less dark in color, and in every respect an improved variety of the race. They were chiefly engaged in eating a large yellow fruit like a gourd, sections of which they divided with their fingers, and ate with rather uncouth voracity, throwing away the rind. A smaller red fruit, shaped like a cucumber, which we had often seen pendant from trees having a broad dark leaf, was also lying in heaps in the centre of several of the festive groups; but the only use they appeared to make of it was sucking its juice, after rolling it between the palms of their hands and nibbling off an end. They seemed eminently happy, and even polite, for we saw, in many instances, individuals sitting nearest these piles of fruit, select the largest and brightest specimens, and throw them archwise across the circle to some opposite friend or associate who extracted the nutriment from those scattered around him, and which were frequently not a few. While thus engaged in their rural banquets, or in social converse, they were always seated with their knees flat upon the turf, and their feet brought evenly together in the form of a triangle. And for some mysterious reason or other this figure seemed to be an especial favorite among them; for we found that every group or social circle arranged itself in this shaped before it dispersed, which was generally done at the signal of an individual who stepped into the centre and brought his hands over his head in an acute angle. At this signal each member of the company extended his arms forward so as to form an acute angle horizontal angle with the extremity of the fingers. But this was not the only proof we had that they were creatures of order and subordination. We had no opportunity of seeing them actually engaged in any work of industry or art; and so far as we could judge, they spent their happy hours in collecting various fruits in the woods, in eating, flying, bathing, and loitering about on the summits of precipices. But although evidently the highest order of animals in this rich valley, they were not its only occupants. Most of the other animals which we had discovered elsewhere, in very distant regions, were collected here; and also at least eight or nine new species of quadrupeds. The most attractive of these was a tall white stag with lofty spreading antlers, black as ebony. We several times saw this elegant creature trot up to the seated parties od the semi-human beings I have described, and browse the herbage close beside them, without the least manifestation of fear on his part or notice on their. The universal state of amity among all classes of lunar creatures, and the apparent absence of every carnivorous or ferocious creatures, gave us the most refined pleasure, and doubly endeared to us this lovely nocturnal companion of our larger, but less favored world."

"With the careful examination of this instructive valley, and a scientific classification of its animal, vegetable, and mineral productions, the astronomers closed their labors for the night; labors rather mental than physical, and oppressive, from the extreme excitement which they naturally induced. A singular circumstance occurred the next day, which three the telescope quite out of use for nearly a week, by which time the Moon could be no longer observed that month. The great lens, which was usually lowered during the day, and placed horizontally, had, it is true, been lowered as usual, but had been inconsiderately left in a perpendicular position. Accordingly, shortly after sunrise the next morning, Dr. Herschel and his assistants, Dr. Grant and Messrs. Drummond and Home, who slept in a bungalow erected a short distance from the observatory circle, were awakened by the loud shouts of some Dutch farmers and domesticated Hottentotts (who were passing with their oxen to agricultural labor), that the "big house" was on fire! Dr. Herschel leaped out of bed from his brief slumbers, and, sure enough, saw his observatory enveloped in a cloud of smoke. Luckily it had been thickly covered, within and without, with a coat of Romanplaster, or it would inevitably have been destroyed will all its invaluable contents; but, as it was, a hole 15 ft in circumference had been burnt completely though the "reflecting chamber," which was attached to the side of the observatory nearest the lens, through the canvass field on which had been exhibited so many wonders that will ever live in the history of mankind, and through the outer wall. So fierce was the concentration of the solat reyas through the gigantic lense, that a clump of trees standing in a line with them was set on fire, and the plaster of the observatory walls, all round the oriface, was vitrified to blue glass. The lens being almost immediately turned, and a brook of water being within a few 100 yards, the first was soon extinguished, but the damage already done was not inconsiderable. The microscope lenses had fortunately been removed for the purpose of being cleaned, but several of the metallic reflectors were so fused as to be rendered useless. Masons and carpenters were procured from Cape Town with all possible dispatch, and in about a week the whole apparatus was again prepared for operation."

"The Moon being now invisible Dr. Herschel directed his inquiries to the primary planets of the system, and first to the planet Saturn. We need not say that this remarkable globe has for many ages been an object of the most ardent astronomical curiosity. The stupendous phenomenon of its double ring having baffled the scrutiny and conjecture of many generations of astronomers, was finally abandoned as inexplicable. It is well known that this planet is stationed in the system 900 millions of miles distant from the sun, and that having the immense diameter of 79,000 miles, it is more than 900 times larger than Earth. The annual motion round the sun os not accomplished in less that 29½ of our years, whilst its diurnal rotation upon its axis is accomplished in 10 h. 16 m., or considerably less than half a terrestrial day. It has not less than seven Moons, the sixth and seventh of which were discovered by the elder Herschel in 1759. It is thwarted by mysterious belts or bands of yellowish tinge, and is surrounded by a double ring—the outer one of which is 204,000 miles in diameter. The outside diameter of the inner ring is 184,000 miles, and the breadth of the outer one being 7200 miles, the space between them is 28,000 miles. The breadth of the inner ring is much greater than that of the other, being 20,000 miles; and its distance from the body of Saturn is more than 30,000. These rings are opaque, but so thin that their edge has not until now been discovered. Sir John Herschel's most interesting discovery with regard to this planet is the demonstrated fact that these two rings are composed of the fragments of two destroyed worlds, formerly belonging to our solar system, and which, on being exploded, were gathered around the immense body of Saturn by the attraction of gravity, and yet kept from falling to its surface by the great centrifugal force created by its extraordinary rapidity on its axis. The inner ring was therefore the first of these destroyed worlds (the former station of which in the system is demonstrated in the argument which we sub-join), which was accordingly carried round by the rotary force, and spread forth in the manner we see. The outer ring is another world exploded in fragments, attracted by the law of gravity as in the former case, and kept from uniting with the inner ring by the centrifugal force of the latter. But the latter, having a slower rotation than the planet, has an inferior centrifugal force, and accordingly the space between the outer and inner ring is ten times less than between the inner ring and the body of Saturn. Having ascertained the mean density of the rings, as compared with the density of the planet, Sir John Herschel has been enabled to effect the following beautiful demonstration. [Which we omit, as too mathematical for popular comprehension."—*Ed. Sun.*]

"Dr. Herschel clearly ascertained that these rings are composed of rocky strata, the skeletons of former globes, lying is a state of wild and ghastly confusion, but not devoid of mountains and seas. The belts across the body of Saturn he has discovered to be the smoke of a number of intense volcanoes, carried in these straight lines by the extreme velocity of the rotary motion. [And these also he has ascertained to be the belt of Jupiter. But the portion of the work which is devoted to this subject, and to the other planets, as also that which describes the astronomer's discoveries among the stars, is comparatively uninteresting to general readers, however highly it might interest others of scientific taste and mathematical acquirements."—*Ed. Sun.*]

"It was not until the new Moon of the month of March, that the weather proved favorable to any continued series of lunar observations; and Dr. Herschel had been so enthusiastically absorbed in demonstrating his brilliant discoveries in the southern constellations, and in constructing tables and catalogues of his new stars, to avail himself of the few clear nights which intervened."

"On one of these, however, Mr. Drummond, myself, and Mr. Holmes, made these discoveries near the Bay of Rainbows, to which I have somewhere briefly alluded. This bay thus fancifully denominated is a part of the northern boundary of the first great ocean which I have lately described, and is marked on the chart with the letter O. The tract of country which we explored on this occasion is numbered 6,5,8,7, in the catalogue, and the chief mountains to which these numbers are attached are severally named Atlas, Hercules, Heraclides Verus, and Heraclides Falsus. Still farther to the north of these os the island circle called Pythagoras, and numbered 1; and yet nearer the meridian line is the mountainous district marked R, and called the Land of Drought, and Q, the Land of Hoar Frost; and certainly the name of the latter, however theoretically bestowed, was not altogether inapplicable, for the tops of its very lofty mountains were evidently covered with snow, though the valleys surrounding them were teeming with the luxuriant fertility of midsummer. But the region which we first particularly inspected was that of Heraclides Falsus (No. 7), in which we found several new specimens of animals, all of which were horned and of a white or grey color, and the remains of three ancient triangular temples which had long been in ruins. We thence traversed the country southeastward, until we arrived at Atlas (No. 6), and it was one of the noble valleys at the foot of this mountain that we found the very superior species of the Vespertilio-homo. In stature they did not exceed those last described, but they were of infinitely greater personal beauty, and appeared in our eyes scarcely less lovely than the general representations of angels by the more imaginative schools of painters. Their social economy seemed to be regulated by laws or ceremonies exactly like those prevailing in the Vale of the Triads, but their works of art were more numerous, and displayed a proficiency of skill quite incredible to all except actual observers. I shall, therefore, let the first detailed account of them appear in Dr. Herschel's authenticated natural history of this planet...."

The real mystery of the Great Moon Hoax concerns the identity of the author. Most claim it was Richard Adams Locke, a Cambridge-educated reporter who, in August 1835, was working for *The Sun*. Locke never publicly admitted to being the author, while rumors persisted that others were involved. Two other men have been noted in connection with the hoax: Jean-Nicolas Nicollet, a French astronomer traveling in America at the time (though he was in Mississippi, not New York, when the Moon-hoax issues appeared), and Lewis Gaylord Clark, editor of *The Knickerbocker*, a literary magazine. However, there is no good evidence to indicate that anyone but Locke was the author of the hoax.

Assuming that Richard A. Locke was the author, his intentions were probably, first, to create a sensational story which would increase sales of *The Sun*, and, second, to ridicule some of the more extravagant astronomical theories that had been published. The fantastic story, though too wild to be true, garnered a healthy following for the newspaper, and even Sir John Herschel found it amusing—until he later had to answer questions from readers who thought the story was serious. The articles where published under the name Dr. Andrew Grant, who described himself as the traveling companion and laborer of Sir John Herschel, but Dr. Grant was fictitious. The story was not discovered to be a hoax until several weeks after its publication, and even then the newspaper did not issue a retraction. Eventually it was announced that the huge telescope used for the observations had caught a glimpse of the Sun, magnified the beams and the observatory caught fire, terminating any further views of the fantastic landscapes on the Moon.

3. How the Moon Formed

Scientists have offered several major theories to account for the origin of the Moon. All have drawbacks, but the favored theory that emerged from the Apollo missions was the Giant Impact Hypothesis (sometimes called the Big Splat). This states that our Moon was created by a collision between Earth and a Mars-sized object some 4.5 billion years ago.

There are a number of variations and alternatives, including captured body, fission, formed together (condensation theory), planetesimal collisions (formed from asteroid-like bodies), and collision theories. All of the theories have been challenged, and none satisfy all questions.

NASA scientist Dr. Robin Brett sums it up best: "It seems much easier to explain the nonexistence of the Moon than its existence."

Giant Impact Hypothesis (GIH)

The collision of two protoplanetary bodies during the early accretional period of Solar System evolution is the most widely accepted explanation for the origin of the Moon. This theory became popular in 1984. It satisfies the orbital conditions of Earth and the Moon and can account for the relatively small metallic core of the Moon. Collisions between planetesimals are now recognized to lead to the growth of planetary bodies. In this framework it is inevitable that large impacts will sometimes occur when the planets are nearly formed. The theory is thought to have originated in the 1940s with Reginald Aldworth Daly, head of the geology department at Harvard (Fig. 3.1).

The colliding body is called Theia, the mother of Selene, the Moon goddess in Greek mythology. The hypothesis requires a collision between a body about 90 % the present size of Earth, and another the diameter of Mars (half of the terrestrial radius and a

© Springer International Publishing Switzerland 2016
V.S. Foster, *Modern Mysteries of the Moon*, Astronomers' Universe,
DOI 10.1007/978-3-319-22120-5_3

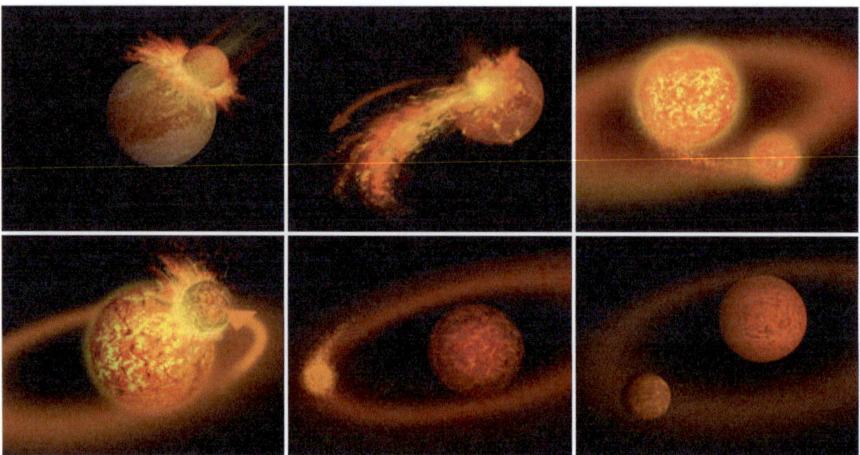

Fɪɢ. **3.1** The giant impact theory suggested that growing smaller planetary body (Mars-sized protoplanet) hit Earth about 4.5 billion years ago, blowing out rocky debris that was captured into orbit around Earth and coalesced into the Moon. NASA illustration

tenth of its mass). This size ratio is needed in order for the resulting system to possess sufficient angular momentum to match the current orbital configuration. Such an impact would have put enough material in orbit around Earth that could accumulate and eventually form the Moon.

Computer simulations show a need for a glancing blow, which would cause a portion of the colliding body to form a long arm of material that then would shear off. The asymmetrical shape of Earth following the collision would then cause this material to settle into an orbit around the main mass. The energy produced by this collision would have been impressive: trillions of tons of material would have been vaporized and melted. The temperature would have risen to 10,000 °C (18,000 °F).

The relatively small iron core of the Moon is explained by Theia's core accreting into Earth's. The lack of volatiles in the lunar samples is also explained in part by the energy of the collision. The energy liberated during the re-accretion of material would have been sufficient to melt a large portion of the Moon, leading to the generation of a magma ocean.

The newly formed Moon orbited 90 % closer than it does today. It became tidally locked with Earth, where one side continually faces toward Earth. The geology of the Moon has since become more independent of Earth. Although this hypothesis explains many aspects of the Earth-Moon system, there are still unresolved problems with it, such as why the Moon's volatile elements were not depleted as expected from such an energetic impact.

Comparisons of lunar and Earth isotopes is another issue. In 2011, the most precise measurement yet of the isotopic signatures of lunar rocks was published. Surprisingly, it showed that the Apollo lunar samples carried an isotopic signature identical to Earth rocks, but different from other Solar System bodies. Since most of the material that went into orbit to form the Moon was thought to come from Theia, this observation was unexpected. In 2007, researchers from Caltech found that the likelihood of Theia having an identical isotopic signature to Earth was very small (<1 %).

An analysis of titanium isotopes in Apollo lunar samples in 2012 revealed that the Moon has the same composition as Earth, which conflicts with the Moon forming far from Earth's orbit. To help explain problems with this, a new theory was published by R. M. Canup in late 2012 which posits that two bodies five times the size of Mars collided, then re-collided, forming a large disc of debris that eventually formed Earth and the Moon.

Released at the same time was another study on the depletion of zinc isotopes on the Moon, which supported the giant impact origin for Earth and the Moon. In 2013, a study was released indicating that water in lunar magma was 'indistinguishable' from carbonaceous chondrites and nearly the same as Earth's, based on the composition of isotopes. Another challenge was issued September 2013, with a growing sense that lunar origins are much more complicated than can be fully explained by the GIH theory.

The Capture Theory

This hypothesis states that Earth captured the Moon. This was popular until the 1980s. Some things in favor of this model include the Moon's size, orbit, and tidal locking. One problem is understanding the capture mechanism. A close encounter with Earth typically results in either collision or altered trajectories. For this hypothesis to work, the primitive Earth would had to have had an extended atmosphere around it, which would have been able to slow the movement of the Moon before it could escape. That may also explain the irregular satellite orbits of Jupiter and Saturn. In addition, this hypothesis has difficulty explaining the identical oxygen isotope ratios of Earth and the Moon.

The Fission Theory

This is the idea that an ancient, rapidly spinning Earth expelled a piece of its mass. This theory was proposed by George Darwin (son of the famous biologist Charles Darwin) in the 1800s and retained some popularity until Apollo. The Austrian geologist Otto Ampherer in 1925 also suggested that the emerging Moon was cause for continental drift.

It was proposed that the Pacific Ocean represented the scar of this event. However, today it is known that the oceanic crust that makes up this ocean basin is relatively young, about 200 million years old or less, whereas the Moon is much older. However, the assumption that the Pacific is not the result of lunar creation does not disprove the fission hypothesis. This hypothesis also cannot account for the angular momentum of the Earth-Moon system, which is the gravitational torque between the Moon and the tidal bulge of Earth that causes the Moon to be constantly shifted to a slightly higher orbit and Earth to be decelerated in its rotation (Fig. 3.2).

The Accretion Theory

The hypothesis of accretion proposes that Earth and the Moon formed together as a double system from the primordial accretion disk of the Solar System. The problem with this hypothesis is that it does not explain the angular momentum of the Earth-Moon system or why the Moon has a relatively small iron core compared to that of Earth (25 % of its radius compared to 50 % for Earth) (Fig. 3.3).

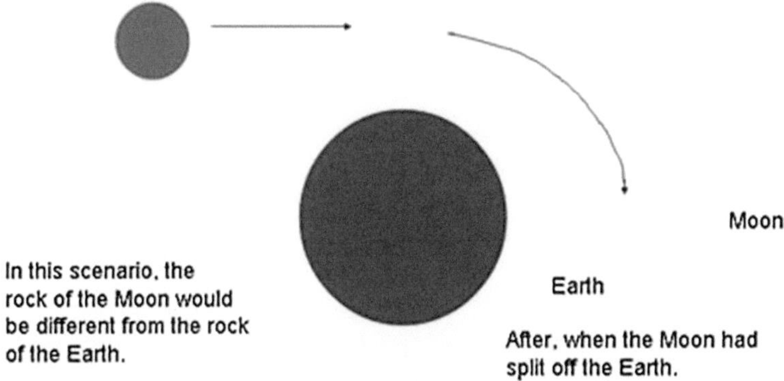

FIG. 3.2 The Capture theory

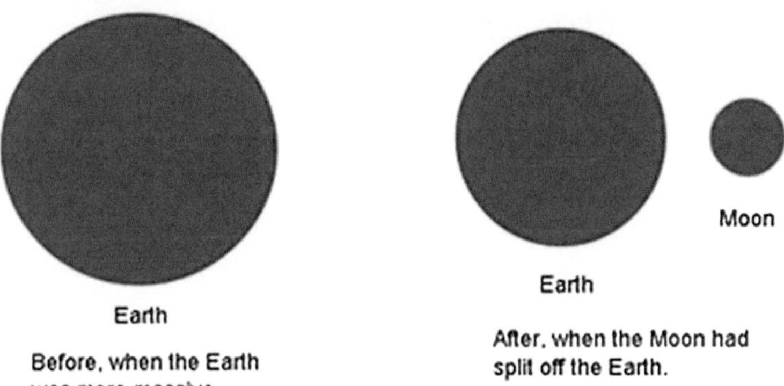

FIG. 3.3 The Fission theory

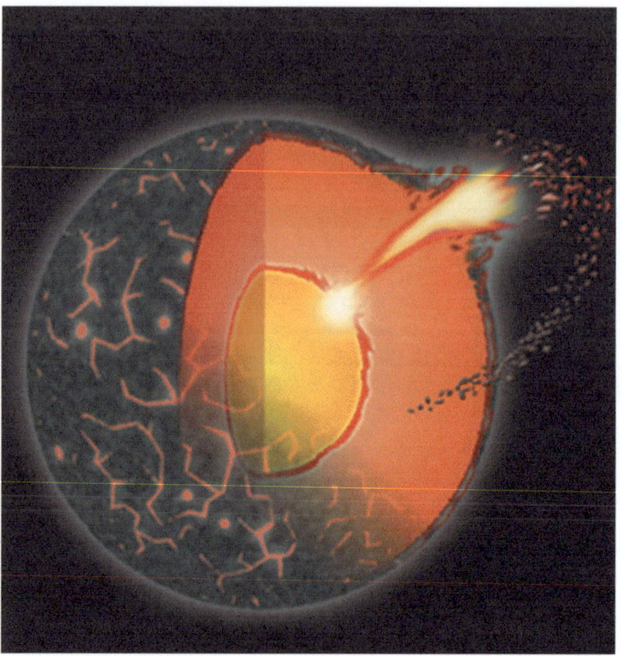

FIG. 3.4 This theory, published in 2010, involves the rapidly spinning Earth having an explosion that originated from a georeactor. This explosion sent a massive body of materials that formed the Moon. This theory easily explains the relation in the compositional similarities between the Moon and the Earth. Though this is a more radical theory compared to the others, it can account for most of the strange formations on the Moon

The Georeactor Explosion Theory

A more radical alternative hypothesis proposes that the Moon may have been formed from the ejected material after the explosion of a georeactor located along the core-mantle boundary at the equatorial plane of the rapidly rotating Earth. A georeactor is a "natural" nuclear reactor—a uranium deposit where a self-sustaining nuclear reaction could occur. This hypothesis could explain the compositional similarities (Fig. 3.4).

In a *Nature* journal study led by Randal Paniello of Washington University, St. Louis a look at lunar chemistry finds fresh evidence confirming a big smash-up as part of the explanation for the Moon. The study examined the zinc seen in the lunar soil and asked how it differed from the same element as seen on Earth, Mars, and asteroids.

In particular, the authors report, the Moon appears conspicuously loaded with a heavier variety of zinc that doesn't vaporize as easily at high temperatures. The results suggest that the Moon formed not from the leftover crust of the Mars-sized planet that hit the young Earth, but from a debris disk of material thrown up by both planets. Volatile elements in that dust ring cooked off before snowballing together to form the Moon.

The idea represents "an entirely new model for Moon formation, in which the impactor hits a rapidly spinning proto-Earth," says planetary scientist Tim Elliott of the United Kingdom's University of Bristol, in a commentary accompanying the study.

"It's nice to see how many aspects and variants of the giant impact idea are still alive and kicking, 38 years after we gave the first paper on the idea, "says planetary scientist William Hartmann of the Planetary Science Institute in Tucson. Hartmann says that an "impact trigger" explanation, where the Moon formed from a debris disk, might be a better name for what happened than a "giant impact," one where the Moon was quickly carved from another planet.

However, new research from geophysical scientist Junjun Zhang and her colleagues at Origins Lab at the University of Chicago suggests that the giant impact hypothesis of the creation of the Moon might be wrong. In comparing titanium isotopes from both the Moon and Earth, the team found that the match is too close to support the theory that the Moon could have been made partly of material from another planet. Instead, it indicates that the Moon's material came from Earth alone.

If two objects had given rise to the Moon, "just like in humans, the Moon would have inherited some of the material from Earth and some of the material from the impactor, approximately half and half," said Nicolas Dauphas, associate professor in geophysical sciences at the University of Chicago and co-author of the study. "What we found is that the child does not look any different compared to the Earth," Dauphas said. "It's a child with only one parent, as far as we can tell."

Researchers selected titanium for their study because the element is highly refractory. This means that titanium tends to remain in a solid or molten state rather than becoming a gas when exposed to tremendous heat. The resistance of titanium isotopes to vaporization makes it less likely that they would become incorporated by the Earth and the developing Moon in equal amounts.

Titanium also contains different isotopic signatures forged in countless stellar explosions that occurred before the Sun's birth. These explosions flung subtly different titanium isotopes into interstellar space. Different objects in the newly forming Solar System gobbled up those isotopes in different ways through collisions, leaving clues that let scientists infer where the solar materials, including the Moon, came from.

Most scientists agree that if a planet had smacked into Earth and the Moon came about as a result, then the Moon ought to be made of some of that other planet as well. Some say the laws of physics suggest it would be somewhere in the neighborhood of 40 %. If that's the case, why don't studies of rocks brought back by the Apollo missions show evidence of this other planet?

However, a new series of measurements of the ratios of oxygen isotopes provides increasing evidence that the Moon formed from the collision of Earth with another large planet-sized astronomical body around 4.5 billion years ago.

These ratios are known to vary throughout the Solar System, but their close similarity between Earth and the Moon conflicted with theoretical models of the collision that indicated that the Moon would form mostly from Theia and thus would be expected to be compositionally different from Earth.

In more recent times a group of German researchers, led by Daniel Herwartz, has used more refined techniques to compare the ratios of $^{17}O/^{16}O$ in lunar samples with those from Earth. The team initially used lunar samples that had arrived on Earth via meteorites, but as these samples had exchanged their isotopes with water from Earth, fresher samples were sought. These were provided by NASA from the *Apollo 11, 12,* and *16* missions; they were found to contain significantly higher levels of $^{17}O/^{16}O$ than their Earthly counterparts.

"The differences are small and difficult to detect, but they are there," said Herwartz. "This means two things: Firstly, we can now be reasonably sure that the giant collision took place. Secondly, it gives us an idea of the geochemistry of Theia. Theia seems to have been similar to what we call E-type chondrites.

If this is true, we can now predict the geochemical and isotopic composition of the Moon because the present Moon is a mixture of Theia and early Earth. The next goal is to find out how much material of Theia is in the Moon."

Most models estimate that the Moon is composed of around 70–90 % material from Theia, with the remaining 10–30 % coming from early Earth. However, some models argue for as little as 8 % Theia in the Moon. Herwartz said that the new data indicate that a 50:50 mixture seems possible, but this needs to be confirmed.

The team used an advanced sample preparation technique before measuring the samples via stable isotope ratio mass spectrometry, which showed a 12 parts per million (±3 ppm) difference in $^{17}O/^{16}O$ ratio between Earth and the Moon.

The Big Splat Gets New Spin

The Moon did indeed coalesce out of tiny bits of pulverized planet blasted into space by a catastrophic collision 4.5 billion years ago, two new studies suggest.

The new research potentially plugs a big hole in the Giant Impact Theory, long the leading explanation for the Moon's formation. Previous versions of the theory held that the Moon formed primarily from pieces of a mysterious Mars-size body that slammed into a proto-Earth, but that presented a problem, because scientists know that the Moon and Earth are made of the same stuff.

The two studies explain how Earth and the Moon came to be geochemical twins. However, they offer differing versions of the enormous smashup that apparently created Earth's natural satellite, giving scientists plenty to chew on going forward.

One of the studies—by Matija Cuk of the SETI (Search for Extraterrestrial Intelligence) Institute in Mountain View, CA, and Sarah Stewart of Harvard—suggests the answer lies in Earth's rotation rate (2010).

If Earth's day had been just 2–3 h long at the time of the impact, Cuk and Stewart calculate, the planet could well have thrown off enough material to form the Moon (which is 1.2 % as massive as Earth).

This rotational speed might sound incredible, and indeed it's close to the threshold beyond which the planet would begin to fly apart. But researchers say the early Solar System was a "shooting gallery" marked by many large impacts, which could have spun planets up to enormous speeds.

Cuk's and Stewart's study, which appears online in the journal *Science in 2011*, also provides a mechanism by which Earth's rotation rate could have slowed over time. After the collision, a gravitational interaction between Earth's orbit around the Sun and the Moon's orbit around Earth could have put the brakes on the planet's super-spin, eventually producing a 24-h day, scientists determined.

Cuk's and Stewart's version of the cosmic smashup posits a roughly Mars-size impactor—a body with 5–10 % of the mass of Earth. However, the other new study—being published in the same issue of *Science*—envisions a collision between two planets in the same weight class. "In this impact, the impactor and the target each contain about 50 % of the [present] Earth's mass," said Robin Canup, of the Southwest Research Institute in Boulder, CO.

"This type of impact has not been advocated for the Earth-Moon before (although a similar type of collision has been invoked for the origin of the Pluto-Charon pair)," Canup added, referring to the largest moon of Pluto. In her computer models, the symmetry of this collision caused the resulting Moon-forming debris disk to be nearly identical in composition to the mantle of the newly enlarged Earth.

Canup's models further predict that such an impact would significantly increase Earth's rotational speed. But that may not be a big issue, since Cuk's and Stewart's work explains how Earth's spin could have slowed over time.

A third study, published in the journal *Nature*, determined that huge amounts of water boiled away during the Moon's birth. This finding, made by examining Moon rocks brought back to Earth by Apollo astronauts, further bolsters the broad outlines of the giant impact theory.

Though the gigantic smashup occurred 4.5 billion years ago, scientists may one day be able to piece together in detail how it all went down, Canup said. "Models of terrestrial planet assembly should be able to evaluate the relative probability of, e.g., the collision I advocate vs. the one proposed by Cuk and Stewart," she said.

The team based the research on two recent studies, one of which Stewart conducted with Matija Cuk of the SETI (Search for Extraterrestrial Intelligence) Institute in Mountain View, CA. That research argued that the Moon is actually a giant merger of bits and pieces of our own planet, partially destroyed by a catastrophic collision with a space body 4.5 billion years ago.

Back then, Earth had a 2- or 3-h day, she said, and the impact made it throw off enough material to coalesce into what became our satellite, making it Earth's geochemical twin. This ultra-rapid spin is one of the important conditions necessary to make the atmospheric loss theory work, Stewart said.

Wandering Gas Giants and Lunar Bombardment

There may have been a dramatic event early in the history of the Solar System—during the intense bombardment of the inner planets and the Moon by planetesimals in a narrow interval of between 3.92 and 3.85 billion years ago called the late heavy bombardment, (also nicknamed the lunar cataclysm). The evidence for this event comes from Apollo lunar samples and lunar meteorites. Although not proven, it makes for an interesting working hypothesis. If correct, what caused it to happen? (Fig. 3.5).

A group of physicists from the Observatoire de la Côte d'Azur (Nice, France), the GEA/OV/Universidade Federal do Rio de Janeiro and Observatório Nacional/MTC (Rio de Janeiro, Brazil), and the Southwest Research Institute (Boulder, CO) conducted a series of studies of the dynamics of the early Solar System. Alessandro Morbidelli, Kleomenis Tsiganis, Rodney Gomes, and Harold Levison simulated the migration of Saturn and Jupiter.

When the orbits of these giant planets reached the special condition of Saturn making one trip around the Sun for every two

Fig. 3.5 Artist's impression of the Moon during the Late Heavy Bombardment (lunar cataclysm) and today. (Credit: NASA)

trips by Jupiter (called the 1:2 resonance), violent gravitational shoves made the orbits of Neptune and Uranus unstable, causing them to migrate rapidly and scatter countless planetesimals throughout the Solar System.

This dramatic event could have happened in a short interval, anywhere from 200 million years to a billion years after planet formation, causing the lunar cataclysm, which would have affected all the inner planets. There are lots of really old lunar rocks. Ferroan anorthosites, which were the first to accumulate from the ocean of magma surrounding the Moon when it formed, crystallized 4.45 billion years ago.

However, many, many rocks formed by melting during huge impact events, which we call impact melt breccias, have ages that fall into a narrow time interval of between 3.92 and 3.85 billion years. This apparent clustering of ages was first noticed in the mid-1970s by Fouad Tera, Dimitri Papanastassiou, and Gerald Wasserburg (Caltech), who concluded that the ages record an intense bombardment of the Moon. They called it the "lunar cataclysm" and proposed that it represented a dramatic increase in the rate of bombardment of the Moon around 3.9 billion years ago. More recent work on lunar samples and lunar meteorites generally confirms that there is a dearth of ages for impact melts older than 3.9 billion years.

The lunar cataclysm is an established, solid idea. Or is it? No, say the voices from the critics' corner. Randy Korotev (Washington University in St. Louis) is skeptical of the whole idea, as was his late colleague Larry Haskin. Korotev thinks we have a hideous sampling problem, and that the Apollo sites were all too close to the Imbrium impact basin. Imbrium is 1300 km in diameter and tossed its continuous ejecta over an area twice that size. (The basalt flows composing Mare Imbrium make up a thin veneer that covers only part of the impact basin.)

They say that all the impact melt breccias we have are associated with the Imbrium impact. No wonder they all have the same age—they were all made by one gigantic event. Most lunar scientists do not agree with this hardnosed interpretation. They point out that many of the samples of impact melts cluster into geochemical groups that have distinctive ages. Although the ages do not vary much from cluster to cluster, they do differ beyond experimental uncertainties. Nevertheless, it is difficult to prove the Imbrium-only hypothesis wrong... and really hard to convince Randy Korotev that he should abandon the idea and embrace the cataclysm interpretation!

The skeptics do have some rock data on their side. A group of feldspar-rich impact melt breccias from the *Apollo 16* landing site have ages between 4.09 and 4.14 billion years, averaging 4.12 billion years. This is substantially older than the narrow cataclysm range. If these ages represent the age of an impact, it shows that impacts certainly took place before 3.9 billion years. And if the ages represent the age of a basin, such as the Nectaris Basin a few 100 km to the east, then it casts great doubt on the cataclysm hypothesis. The feldspar-rich composition of these rocks is consistent with remote-sensing observations of the lunar highlands surrounding Nectaris. However, the impact melted portion in the rocks consists of very small grains, implying rapid cooling.

As the late Graham Ryder (Lunar and Planetary Institute) argued, this means that the embedded feldspar fragments, which come from ancient rocks, might not have been heated enough by the impact, so they retain an isotopic memory of the older crustal igneous rocks. In other words, the formation of the impact melt breccia did not reset the clock to the time of the impact event.

Bill Hartmann (Planetary Science Institute, Tucson) has had completely different reasons for being skeptical. Driven by the lack of a sound mechanism for a cataclysm, he always preferred a declining impact rate from the time of lunar formation until 3.8 billion years ago. He suggests that this spread-out early bombardment continually pulverized the upper reaches of the lunar crust, systematically removing impact melt breccias that formed early. The only survivors as large fragments are the most recent ones, giving us the false impression that the ages of impact melts cluster between 3.85 and 3.92 billion years. He calls it the stone-wall effect.

The argument against Hartmann's interpretation is that we have samples of lava flows that have ages up to 4.25 billion years. If these surface rocks survived the declining early bombardment, surely some impact melt breccias would have, too. Ironically, if the feldspar-rich *Apollo 16* impact melt breccias are shown to be over 4.1 billion years old, it would simultaneously weaken the cataclysm idea and Hartmann's stone-wall hypothesis against the cataclysm interpretation!

It is important to test the cataclysm hypothesis with additional lunar samples, as discussed below. For now, let's assume that there was a lunar cataclysm. What could have caused it? We normally think of the orbits of the planets as unchanging and reliable. Sure, stray asteroids and comets can come close to Earth and even hit it, maybe causing extinctions of numerous species. But not the planets. They stay put. We can rely on them. Well, it turns out that early in the history of the Solar System, the planets may have roamed, especially the giants Jupiter, Saturn, Uranus, and Neptune. Jupiter seems to have migrated in towards the Sun, while the others wandered away from the Sun. When their orbits reached certain simple relationships to each other, serious gravitational pushing and pulling happened that violently destabilized the orbits of Uranus and Neptune, thus flinging millions of leftovers from planet formation (asteroid-to-Moon-sized planetesimals) throughout the Solar System. Planets moved and craters formed, possibly in a dramatic, chaotic, messy moment of geologic history.

This is the story of the early Solar System being developed by physicists and astronomers, including Morbidelli, Tsiganis, Gomes, and Levison. Current theory says that migrating planets are a natural consequence of planet formation. Such migration would be accompanied by changes in the inclination and eccentricity of the outer planets, increasing from nice, co-planar circular orbits to those that are inclined to the ecliptic (Earth's orbital plane) and not exactly circular (they follow elliptical paths).

After the planets were constructed, Levison and his colleagues say, the Solar System still teemed with the detritus of planet formation—planetesimal that had not accreted to a planet. These stunted planets were indiscriminately hurled around by gravitational interactions with the outer planets. These interactions, called dynamical friction, caused changes in the orbits of the outer planets, too. Jupiter moved inward toward the Sun as Saturn, Neptune, and Uranus migrated outward. As Saturn drifted outward it eventually reached a resonance with Jupiter, in which it orbited the Sun once for every two orbits of Jupiter around the Sun. This is the 1:2 resonance mentioned earlier.

The special timing of the orbits of these two giant planets caused their gravitational interactions to pump up the motions of Neptune and Uranus. Tsiganis and his colleagues calculate, for example, that the orbits of Neptune and Uranus might have become highly elliptical, and Neptune's orbital distance from the

Sun doubled, sending it into a zone peppered with planetesimals. The planetesimals were scattered by Neptune's substantial gravity field, sending them all over the place, including into the inner Solar System to bombard the rocky planets and the Moon. The situation was so dynamic that Neptune, which had started closer to the Sun than Uranus, ended up farther away.

There is a timing problem. The lunar cataclysm (if this is the valid interpretation of the ages of lunar impact breccias) took place between about 3.92 and 3.85 billion years ago. This means that Saturn would have had to have moved into the 1:2 resonance with Jupiter 600–700 million years after the formation of the Solar System. What mechanism could delay migration of the giant planets for so long? That's the central focus of the paper by Gomes. He points out that the time at which Jupiter and Saturn reached their 1:2 resonance depended on three major factors. One, logically enough, was their distance from the resonance position. The second was the mass of materials in the planetesimal disk, particularly near its inner edge. The third factor was the relative location of the inner edge of the disk and the outermost ice giant (Neptune or Uranus).

The farther from the Sun it is, the longer it takes for Jupiter and Saturn to reach their planet-scattering 1:2 resonance. At slightly more than 15 astronomical units (AU), the time is in the right range of a few hundred million years. By varying the other parameters, Gomes and his coworkers find that Jupiter and Saturn could have reached their 1:2 resonance between 200 million and 1.1 billion years after formation of the planets.

Although not precise, this shows that the resonance mechanism is reasonable for producing the late heavy bombardment. In fact, if the lunar cataclysm is proven, perhaps its duration and age can be used to set limits on the variables that affect the timing of migration. The simulations suggest that the impactors on the Moon would be from both comets and asteroids, but the exact percentage of each is highly uncertain. However, the dynamicists are sure that there would be a lot of comet impacts, adding on the order of 8×10^{21} g of cometary material to the Moon.

It is curious that we do not see evidence of this influx of icy objects in lunar samples, all of which are bone-dry. Assuming that there was a late heavy bombardment, it must have had dramatic effects on all the planets. The record is obscure on Earth, although the maximum age for Earth rocks is about 3.8 billion years, in the right range. The late Graham Ryder suggested that early bombardment of Earth provided environments in which life could arise, although some astrobiologists have suggested that the bombardment would have sterilized Earth, thus requiring life to begin multiple times.

The surface of Mercury is battered, like that of the Moon. (Venus has been resurfaced by volcanism, erasing all evidence for the early bombardment.) All interesting possibilities, but the effects of the late heavy bombardment may be most evident on Mars. Besides forming thousands of craters larger than 20 km in diameter and perhaps a hundred basins larger than 300 km, the effects of a cataclysmic bombardment of Mars may be widespread. They could include mixing of diverse rock types, as we see in samples collected from the lunar highlands. Formation of large basins might have erased the record of an early magnetic field in Mars, now recorded only in certain regions of the highlands. Impacts would have formed large amounts of impact melt and fragmented breccias, making it difficult to find large expanses of ancient rocks.

Perhaps most interesting, a cataclysmic bombardment of Mars with icy comets would have added a substantial amount of water to the crust. This might have triggered the early wet period in the history of Mars. Each impact might have caused rainy periods, as proposed by Teresa Sigura (University of Colorado, Boulder), leading to intense erosion of the highlands and deposition into craters that are mostly filled with sediment.

Could the early wet period on Mars have been caused by the migration of Jupiter and Saturn? The idea of a short, but strong, spike in the rate of impacts 3.9 billion years ago is an extremely important concept. It is not proven, as Joe Hahn (Saint Mary's University, Halifax) points out in a clear and concise summary of the dynamical calculations. He says that the good agreement with the age of lunar basin formation and the explanation for the eccentricities and inclinations of the outer planets, although very interesting, is not proof that the simulations are correct. Perhaps the eccentricities and inclinations were pumped up by the presence of other large protoplanets that eventually wound up inside a giant planet or were ejected from the Solar System.

What happened to all the water in the comets that would have whacked into the Moon? How could all of it escape? If it did escape, did it also escape from Mars? More important for the Moon, did all those basins really form in a short time around 3.85 billion years ago?

Cosmochemists intend to test this idea. We can use existing Apollo samples, doing more detailed studies of those ambiguous *Apollo 16* Group 4 impact melt breccias and of more fragments of impact melts inside lunar meteorites. Even better, we can date samples returned by robotic missions to lunar basins.

One favorite among cosmochemists is the oldest basin on the Moon, the South Pole-Aitken Basin. The samples could be returned on a robotic mission or by astronauts when humans return to the Moon, an event scheduled for somewhere around 2018. Settlement of the Moon, a central theme of the former national Vision for Space Exploration, would have enabled detailed studies by astronauts living on the Moon of the ages of many lunar basins, including Nectaris and Crisium. It is critical to test the idea of the lunar cataclysm. If proven, then the calculations done by Morbidelli, Tsiganis, Gomes, and Levison are more likely to be correct, and their portrait of processes operating during planet formation may be accurate.

This portrait of planetary migration explains the late heavy bombardment, the orbital parameters of the outer planets, the present orbital distribution of main-belt asteroids, and the presence of Jupiter's Trojan asteroids. If there was no cataclysm, scientists will have to modify their theories, perhaps drastically. As often happens in planetary science, cosmochemists and astronomers are partners trying to unravel the secrets of the early Solar System.

Lunar Paradox

Over the past decades scientists have simulated the formation process and reproduced many of the properties of the Earth-Moon system; however, these simulations have also given rise to a problem known as the Lunar Paradox, that is, that the Moon appears to be made up of material that would not be expected if the current collision theory is correct.

A recent study published in the journal *Icarus* proposes a new perspective on the theory in answer to the paradox. If current theories are to be believed, analyses of the various simulations of the Earth-Theia collision predict that the Moon is mostly made up of material from Theia. However, materials from both Earth and the Moon, show remarkable similarities. In fact, elements found on the Moon show identical isotopic properties to those found on Earth.

Given that it is very unlikely that both Theia and Earth had identical isotopic compositions (as all other known Solar System bodies, except the Moon, appear to be different) this paradox cast doubt over the dominant theory for the Moon formation. Moreover, for some elements, like silicon, the isotopic composition is the result of internal processes, related to the size of the parent body. Given Theia was smaller than Earth, its silicon isotope composition should have definitely been different from that of Earth's mantle.

A group of researchers from the University of Bern, Switzerland, have now made a significant breakthrough in the story of the formation of the Moon, suggesting an answer to this Lunar Paradox. They explored a different geometry of collisions than previously simulated, also considering new impact configurations such as the so-called hit-and-run collisions, where a significant amount of material is lost into space on orbits unbound to Earth.

"Our model considers new impact parameters, which were never tested before. Besides the implications for the Earth-Moon system itself, the considerably higher impact velocity opens up new possibilities for the origin of the impactor and therefore also for models of terrestrial planet formation," explains lead author of the study, Andreas Reufer.

"While none of the simulations presented in their research provides a perfect match for the constraints from the actual Earth-Moon-system, several do come close," adds Alessandro Morbidelli, one of *Icarus*'s editors. "This work, therefore, suggests that a future exhaustive exploration of the vast collisional parameter space may finally lead to the long-searched solution of the lunar paradox."

Moon's Birth Is Rare

The next time you take a moonlit stroll, or admire a full, bright-white Moon looming in the night sky, you might count yourself lucky. New observations from NASA's Spitzer Space Telescope (2011) suggest that moons like Earth's—that formed out of tremendous collisions—are uncommon in the universe, arising at most in only 5–10 % of planetary systems.

"When a moon forms from a violent collision, dust should be blasted everywhere," said Nadya Gorlova of the University of Florida, Gainesville, lead author of a study appearing in the *Astrophysical Journal*. "If there were lots of moons forming, we would have seen dust around lots of stars—but we didn't."

It's hard to imagine Earth without a moon. Our familiar white orb has long been the subject of art, myth, and poetry. Wolves howl at it, and humans have left footprints in its soil. Life itself might have evolved from the ocean to land thanks to tides induced by the Moon's gravity.

Scientists believe the Moon arose about 30–50 million years after our Sun was born, and after our rocky planets had begun to take shape. A body as big as Mars is thought to have smacked into our infant Earth, breaking off a piece of its mantle. Some of the resulting debris fell into orbit around Earth, eventually coalescing into the Moon we see today. The other moons in our Solar System either formed simultaneously with their planet or were captured by their planet's gravity.

Gorlova and her colleagues looked for the dusty signs of similar smash-ups around 400 stars that are all about 30 million years old—roughly the age of our Sun when Earth's Moon formed. They found that only 1 out of the 400 stars is immersed in the tell-tale dust. Taking into consideration the amount of time the dust should stick around, and the age range at which moon-forming collisions can occur, the scientists then calculated the probability of a Solar System making a satellite like Earth's to be at most 5–10 %.

"We don't know that the collision we witnessed around the one star is definitely going to produce a moon, so moon-forming events could be much less frequent than our calculation suggests," said George Rieke of the University of Arizona, Tucson, a co-author of the study.

In addition, the observations tell astronomers that the planet-building process itself winds down by 30 million years after a star is born. Like our Moon, rocky planets are built up through messy collisions that spray dust all around. Current thinking holds that this process lasts from about 10–50 million years after a star forms. The fact that Gorlova and her team found only 1 star out of 400 with collision-generated dust indicates that the 30-million-year-old stars in the study have, for the most part, finished making their planets.

"Astronomers have observed young stars with dust swirling around them for more than 20 years now," said Gorlova. "But those stars are usually so young that their dust could be left over from the planet-formation process. The star we have found is older, at the same age our Sun was when it had finished making planets and the Earth-Moon system had just formed in a collision."

For Moon lovers, the news isn't all bad. For one thing, moons can form in different ways. And, even though the majority of rocky planets in the universe might not have moons like Earth's, astronomers believe there are billions of rocky planets out there. Five to 10 % of billions is still a lot of moons.

Moon Beams Used to Chemically Map Moon

To determine how a planetary body formed and evolved, we must determine the chemical compositions of distinctive geologic regions on it. It is never possible to obtain enough samples of a planet to do this job thoroughly, so planetary scientists have searched for ways of determining chemical compositions from orbit, which would allow chemical mapping of the entire surface. A team at the University of Hawai'i has developed a method to determine the amount of titanium and iron on the lunar surface from the amount of sunlight reflected at different wavelengths.

Most recently (2012), David Blewett, Paul Lucey, B. Ray Hawke (University of Hawai'i), and Bradley Jolliff (Washington University in St. Louis) have used Apollo rock samples to carefully calibrate the technique, allowing surprisingly accurate measurements of iron and titanium. These two elements are especially useful in understanding the origin and geological evolution of the Moon.

You might be wondering, what can you learn from only two elements? It turns out that iron and titanium are quite informative. Titanium is an important probe of the compositions of magmas because it does not enter most minerals; instead it builds up in a magma during crystallization, so its abundance is a measure of the amount of crystallization that took place before a given rock formed in a large body of magma.

Most important, the two elements allow us to distinguish among the known types of lunar rocks, and titanium forms the basis for classifying the basalts that make up the lunar maria. The main types of lunar rocks can be distinguished from one another by their iron and titanium concentrations. (In chemical analyses, these elements are expressed as oxides, FeO and TiO_2, because they are chemically bonded to oxygen inside minerals.) Rocks from the highlands, the light-colored, rough, mountainous areas of the Moon, contain less FeO than do the maria, which are covered with dark lava flows. In fact, different groups within highland and mare rocks can be identified with just these two elements.

There have been two recent missions to the Moon. The *Galileo* spacecraft flew by twice (in 1990 and 1992) as it was getting gravity assists for its trip to Jupiter (where it is now), and the *Clementine* spacecraft was in orbit for a few months in 1994. Both took pictures using special filters that allowed only certain wavelengths to pass through them. This produced images in several wavelengths, and it is the ratios of these wavelengths that allowed Blewett and Lucey to derive the elemental abundances. The *Clementine* mission obtained nearly complete global coverage at excellent spatial resolution; each pixel was about the size of a football field or two.

Deriving the FeO and TiO_2 abundances involves plotting the intensity (called the reflectance) of light at one wavelength against the ratio of the reflectances at two wavelengths. This allows separation of the effects of "space weathering," which darkens the surface, from the effect of composition, which also darkens the surface. Without separation, it is difficult to distinguish one effect from the other. Space weathering is caused on the Moon by the combination of hydrogen being implanted by the solar wind (a stream of particles constantly emitted from the Sun) and impacts by tiny meteorites. The impacts melt some of the lunar regolith (the impact-produced powdery layer on the Moon), while the hydrogen reacts with FeO to produce minute droplets of metallic iron.

Blewett and his coworkers assembled a database of chemical analyses of lunar regolith samples, and then used *Clementine* spectral data to determine the parameters used to calculate the FeO and TiO_2 concentrations. They plotted these parameters against the actual concentrations, as determined from the returned samples. To optimize the data, they adjusted the pivot points on their spectral diagrams to improve the correlation of the parameter with actual FeO or TiO_2 concentration. The resulting correlations are excellent, especially for FeO. The scatter about the lines indicate an uncertainty in the FeO measurement of about 1 wt% and an uncertainty in TiO_2 of 1.1 wt%. The uncertainties mean that measurements when either FeO or TiO_2 are relatively low, say less than 2 wt%, are not very accurate. For example, an apparent FeO concentration of 1 wt% could correspond to an actual FeO content of between 0 and 2 wt%. The scatter for TiO_2 is greater at concentrations less than a few percent.

Lucey and Blewett believe that this might indicate that the variation of the titanium parameter with concentration may not be linear; instead, the correlation may be fitted better by a curved line, rather than a straight one. Why did Blewett, Lucey, and their associates go to all this trouble? The answer is simple: the FeO and TiO_2 contents of the lunar surface contain clues to how the Moon formed, the extent of its initial melting, the compositions of its lava flows, the mechanisms of crater formation, and other interesting facets of lunar history.

Lucey and Blewett are applying their compositional tool to these problems, but most of those results are not yet published. Most of the far side and the highlands portions of the near side are relatively low in FeO. Lucey and coworkers argued in the 1995 paper that this suggested fairly abundant aluminum in the Moon (in lunar rocks low iron correlates with high aluminum, that is, rocks with lower iron tend to have higher aluminum). In fact, they used their measurements to infer that the Moon contains more aluminum than does Earth, ruling out the fission and binary-planet hypotheses for the Moon's origin.

In fission, the primitive Earth spins fast, and a blob is flung off, forming the Moon. This theory implies that the rocky parts of Earth and the Moon have the same chemical compositions. In the binary-planet hypothesis, the Moon forms in orbit around the growing Earth, as a two-object system from the start. Since both growing bodies would be fed by the same material, this idea also implies that the composition of Earth and the Moon are the same.

Still in the running is the idea that the Moon formed as the result of an impact with the growing Earth of an object the size of Mars or larger, the so-called giant-impact hypothesis. In this idea, most of the Moon is formed from chunks of the huge projectile, which is likely to have a different composition than Earth. The idea of the magma ocean is a central theme in lunar science, and has been applied to other planets and some asteroids, but it is not a proven theory.

How do we know that all the highlands are enriched in plagioclase (hence high in aluminum and low in FeO)? After all, we have samples from only a few places. Now, in a way, we have samples from the entire Moon, in the form of the Clementine data. The FeO data shows that most of the highlands are, as the magma ocean hypothesis predicts, low in FeO, and hence probably high in aluminum. So, it is likely that when the Moon formed it was surrounded by an ocean of magma, and the existence of that huge magma body may bear on the details of how the Moon was assembled after the giant impact that formed it.

All that, with more to come, figured out from some samples of lunar dirt and the light reflected from the Moon!

"The Moon is a source of radiation," said Boston University researcher Harlan Spence, the lead scientist for LRO's cosmic ray telescope. "This was a bit unexpected."

Although the Moon blocks galactic cosmic rays to some extent, the hazards posed by the secondary radiation showers counter the shielding effects, Spence told the American Geophysical Union. Overall, future lunar travelers face a radiation dose 30–40 % higher than originally expected, Spence said.

Galactic cosmic rays, believed to be caused by supernova explosions, are fast-moving electrically charged electrons and atomic nuclei. They are found throughout the Solar System, though their numbers are particularly high at the present time due to an unusually quiet period of solar activity.

The lack of solar magnetic fields and reduced solar wind pressure means cosmic rays have been able to more easily penetrate the inner Solar System. "We are in a period when the radiation risks are elevated, but still tolerable," Spence said, adding that the levels were about what an X-ray technician or uranium miner might normally experience in a year.

The Sun cycles through periods of high and low magnetic activity about every 11 years. The most recent period of maximum activity, characterized by an increasing number of sunspots and other disruptions of the Sun's surface, was in 2013.

LRO found about twice as many of the highest energy cosmic rays—those that could punch through the telescope itself—as expected. NASA plans to use the information to design shelters and operating procedures for future human excursions to the Moon.

LRO's primary mission was to scout for landing spots, find potential resources and characterize radiation and other hazards. "The work being done in heliophysics areas is important to keeping astronauts safe," noted the late Michael Wargo, who oversaw NASA's lunar research programs before his death in 2013. "One of the holy grails was to be able to predict the sun's activities and give an 'all clear' on how many days astronauts could be on (spacewalks)."

To help accomplish this, LRO was equipped with the Cosmic Ray Telescope for the Effects of Radiation (CRaTER) detector. Its purpose was to detect radiation in the space around the Moon where the shielding effects of Earth's atmosphere and magnetic field are. It also had a scanner that is covered by a special plastic that reacts to radiation in the same way that human muscle tissue does, allowing researchers to observe the effects of deep space radiation on susceptible bone marrow. The data from the instruments was to provide critical information on the radiation hazards that will be faced by astronauts on extended missions to deep space such as those to Mars (Fig. 3.6).

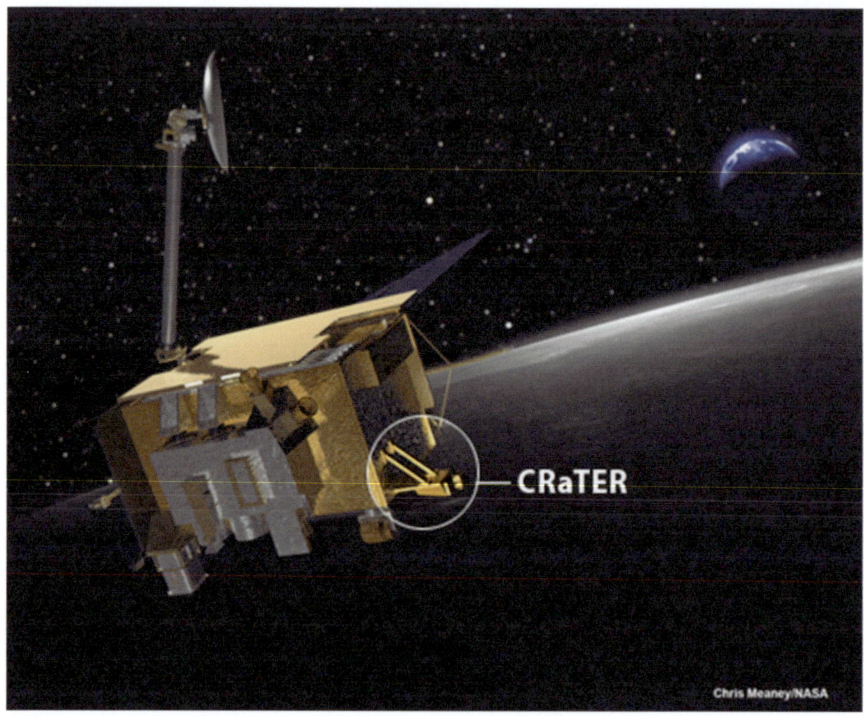

Fɪɢ. **3.6** Pictured is the LRO's science payload of seven instruments to gather data useful to further explore the Moon. Among those instruments is the CRaTER cosmic ray telescope to study the effects of radiation. CRaTER characterizes the global lunar radiation environment and its biological impacts by measuring galactic and solar cosmic ray radiation behind a "human tissue-equivalent" plastic

"These data are a fundamental reference for the radiation hazards in near Earth 'geospace' out to Mars and other regions of our sun's vast heliosphere," said CRaTER principal investigator Nathan Schwadron of the University of New Hampshire Institute for the Study of Earth, Oceans, and Space (EOS).

The space environment poses significant risks to both humans and satellites due to harmful radiation from galactic cosmic rays and solar energetic particles that can easily penetrate typical shielding and damage electronics. When this radiation impacts biological cells, it can cause cancer.

Before CRaTER's long-term radiation measurements were derived using a material called "tissue-equivalent plastic" (TEP)—a stand-in for human muscle capable of gauging radiation dosage—those hazards were not sufficiently well characterized to determine if long missions outside low-Earth orbit can be accomplished with acceptable risk.

Two pieces of TEP, roughly 2 in. long and 1 in. thick, are separated by silicon radiation detectors. The TEP-detector combo measures how much radiation may actually reach human organs, which may be less than the amount that reaches the spacecraft. Data gathered by LRO show lighter materials such as plastics provide effective shielding against the radiation hazards faced by astronauts during extended space travel. The finding could help reduce health risks to humans on future missions into deep space.

Aluminum has always been the primary material in spacecraft construction, but it provides relatively little protection against high-energy cosmic rays and can add so much mass to spacecraft that they become cost-prohibitive to launch.

CRaTER's seminal measurements now provide quantified, radiation hazard data from lunar orbit and can be used to calculate radiation dosage from deep space down to airline altitudes. The data will be crucial in developing techniques for shielding against space-based radiation dosage. The measurements have also played a vital role in University of New Hampshire (UNH) space scientists' efforts to develop both the first Web-based tool for predicting and forecasting the radiation environment in near-Earth, lunar, and Martian space environments and a space radiation detector that possesses unprecedented performance capabilities.

The near real-time prediction/forecasting tool known as PREDICCS integrates for the first time numerical models of space radiation and a host of real-time measurements being made by satellites currently in space. It provides updates of the radiation environment on an hourly basis and archives the data weekly, monthly, and yearly with an historical record that provides a clear picture of when a safe radiation dose limit is reached for skin or blood-forming organs, for example. CRaTER offers an opportunity to test the capability of PREDICCS to accurately describe the lunar radiation environment.

Equipment also included a detector developed at UNH, known as DoSEN, short for Dose Spectra from Energetic Particles and Neutrons. This instrument measures and calculates the absorbed dose in matter and tissue resulting from the exposure to indirect and direct ionizing radiation, which can change cells at the atomic level and lead to irreparable damage. Schwadron is lead scientist for both the PREDICCS and the DoSEN project.

Says Harlan Spence, CRaTER deputy lead scientist and director of EOS, "Until now, people have not had the "eyes" necessary to see this particular population of particles. With CRaTER, we just happen to have the right focus to make these discoveries."

Results from CRaTER indicate that while high-energy cosmic rays from deep space known as HZE particles make up only about 1 % of the radiation in the lunar environment, they carry about 50 % of the total energy from radiation.

Radiation in deep space comes from cosmic rays, from the solar wind, and from solar energetic particles emanated during a solar storm. Particles from these sources rocket through space. Many can pass right through matter, such as our bodies. So-called ionizing radiation knocks electrons off of atoms within our bodies, creating highly reactive ions. Within Earth's protective atmosphere and magnetic field, we receive low doses of background radiation every day. The radiation hazards astronauts face are serious yet manageable, thanks to research endeavors such as the CRaTER instrument.

Larry Townsend of the University of Tennessee, who was a co-investigator on the CRaTER team, notes that CRaTER's observations came at a time when solar activity, and hence the solar wind, had been unusually quiet. The solar wind disperses some galactic cosmic rays, but in the current solar lull, more of these rays are able to bombard Earth and the Moon. "They're lower-level exposures," Townsend said, of galactic cosmic rays, "but they're damaging in the sense that the particles are highly charged and heavy, and they create a lot of damage when they're going through the body."

This radiation comes from the partial reflection, also called an albedo, of galactic cosmic rays off the Moon's surface. Galactic cosmic ray protons penetrate as much as a meter (about 3.2 ft) into the lunar surface, bombarding the material within and creating a spray of secondary radiation and a mix of high-energy particles that flies back out into space. This galactic cosmic ray albedo, which may interact differently with various chemical structures, could provide another method to remotely map the minerals present in the Moon's surface.

CRaTER directly measured the proton component of the Moon's radiation albedo for the first time, said Spence. The TEP radiation detector measures various components of radiation separately, which enables CRaTER to, in Spence's words, "unfold" the energy spectrum of the radiation albedo.

This result, he said, illustrates the value of combining exploration and science in spaceflight. "If we had been on a different science-oriented mission, we probably would've developed a different instrument," Spence said. "In fact, we probably never would have flown TEP."

4. Observing the Moon and Its Mystery Spots

Of all the celestial sights that pass across the sky, none is more inspiring or universally appealing than our planet's lone natural satellite, the Moon. Remember the rush of excitement that you felt when you first peered at the rugged lunar surface through a telescope or binoculars? (If you haven't, you'll be amazed.) The first view of its broad plains, coarse mountain ranges, deep valleys, and countless craters is a memory long cherished by stargazers everywhere.

A New Sight Every Night

Since the Moon orbits our planet in the same time that it takes to rotate once on its own axis, one side of the Moon perpetually faces Earth. Though the face may be the same, its appearance changes dramatically during its 27.3-day orbital period, as sunlight strikes it from different angles as seen from our standpoint. Due to the sunlight's changing angle, the Moon presents a slightly different perspective every night as it passes from phase to phase. No other object in the sky holds that distinction. (Note that it is actually 29.5 days from new Moon to new Moon; the added time is due to Earth's motion around the Sun.)

The Moon is the ideal target for all amateur astronomers. It is bright and large enough to show amazing surface detail, regardless of the type or size of telescopic equipment, and can be viewed just as successfully from the center of a city as from the rural countryside. But bear in mind that some phases are more conducive to Moon watching than others.

© Springer International Publishing Switzerland 2016

V.S. Foster, *Modern Mysteries of the Moon*, Astronomers' Universe,

DOI 10.1007/978-3-319-22120-5_4

The Best Times to Observe

Perhaps the most widespread belief is that the full Moon phase is the best for viewing, but nothing could be further from the truth. Since the Sun is shining directly on the Earth-facing side of the Moon at this phase, there are no shadows to give the lunar surface texture and relief. In addition, the Full Moon is so bright that it can overwhelm the observer's eye. Although no permanent eye damage will result, the Full Moon is uncomfortable to look at even with the naked eye. Instead, the best time to view the waxing Moon is a few nights after New Moon (when the Moon is a thin crescent), up until two or three nights after first quarter. (First quarter is when half of the visible disk is illuminated.) The waning Moon puts on its best show from just before Last Quarter to the New Moon phase. These phases show finer detail because of the Sun's lower elevation in the lunar sky.

No matter what the phase of the Moon is, the view is almost always better through a lunar filter. It screws into the barrel of a telescope eyepiece and cuts the bright glare, making for more comfortable observing and bringing out more surface detail. Some lunar filters, called variable polarizing filters, act something like a dimmer switch, permitting adjustment of the brightness to your liking.

Notable Surface Features

The Moon is dominated by large, flat plains known as maria; the singular is mare (meaning "sea"), which is pronounced (MAH-ray). Maria were first thought to be large bodies of water. In reality, the maria are ancient basins flooded by long-solidified lava created some three billion years ago when the Moon was still volcanically active. All are relatively free of craters except for a few scars from impacts that have occurred since. Their romantic-sounding names, such as the Sea of Crises, Sea of Fertility, Sea of Serenity, Ocean of Storms, and the Sea of Tranquility, are believed to date back to the mid-seventeenth century.

Surrounding the maria are the lunar highlands, dominated by nearly uncountable craters that measure up to several hundred miles across. Most are believed to have been created when debris from the formation of the solar system collided with the young Moon, leaving a permanent record of the barrage on its surface. Some of the more spectacular lunar craters include Tycho, Copernicus, Kepler, Clavius, Plato, and Archimedes, all named for figures of historical stature. Tycho, Copernicus, and Kepler are especially noteworthy, as each displays a broad pattern of bright rays radiating outward. These are particularly impressive during the Moon's gibbous phases (between Quarter and Full), when the Sun appears high in the lunar sky. The Moon also has several noteworthy mountain ranges, such as the Alps and Apennines, as well as straight cliffs, towering ridges, broad valleys, and small, sinuous rilles.

The greatest amount of detail is visible along the Moon's terminator, the line separating the lighted area of the lunar disk from the darkened portion. It is here that the Sun's light strikes the Moon at the narrowest angle. This casts the longest shadows, increasing contrast of lunar features and showing the greatest three-dimensional relief. Sometimes you will notice a bright "island" surrounded by darkness on the dark side of the terminator. That's a high peak, tall enough to still catch the light of the setting Sun, while the lower terrain around it does not.

An amazing world, our Moon, it is so rich in detail and so easy to see. What follows are descriptions of many places on the Moon that still puzzle lunar observers:

Plato

The crater Plato has long been a mystery to observers. The flat floor of the crater has a relatively low albedo, making it appear darker than the surrounding rugged terrain. The floor appears to be free of significant impact craters and lacks a central peak. However there are a several small craterlets scattered across the floor (Fig. 4.1).

The view through a telescope is especially intriguing because of the irregularity of Plato's rim, as shown dramatically by variations in the lengths of shadows cast onto its floor.

Plato is the lava-filled remains of a lunar impact crater. It is located on the northeastern shore of the Mare Imbrium, at the western extremity of the Montes Alpes mountain range. In the mare to

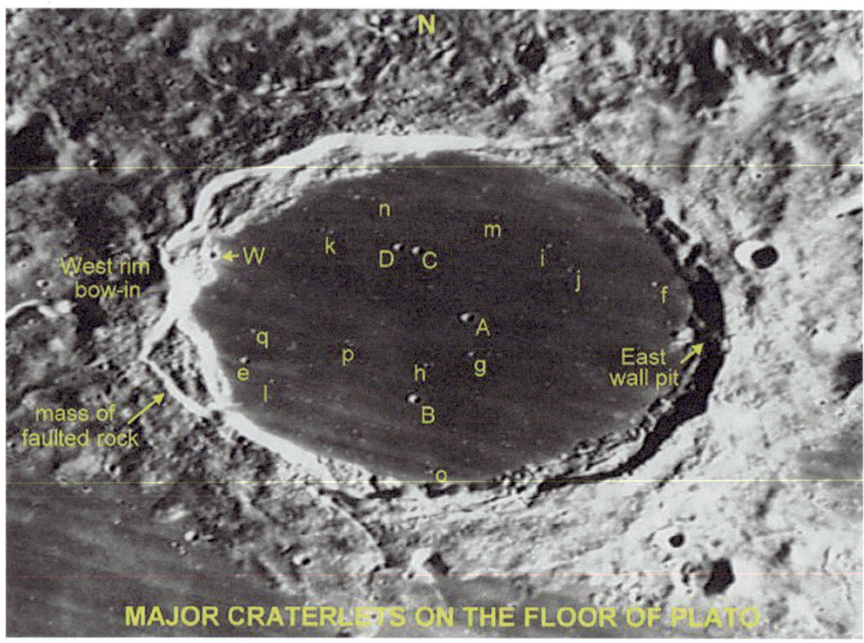

FIG. 4.1 The Big Four. Although many amateurs rarely seem to see very much on the apparently smooth, dark floor of Plato, the above craterlets are the ones most often reported by those lucky enough to get really good seeing. "A" is the easiest of this group due to its fairly prominent ramparts, and has been reported in a 4–5 in. aperture, although the unresolved "bump" of craterlet-A's ramparts is visible in only a 3.1 in. Three or four of these craterlets can sometimes be observed under low Sun angle and in good to excellent seeing in apertures 6 in. and *larger*. These four can sometimes be "detected" as very tiny *white spots* in 3- to 5-in. 'scopes during the full Moon, although to show them all as true pits often requires a 7- or 8-in. aperture. The "East Wall Pit" is a much larger irregular feature (4 miles across) that often hides in the shadow of the eastern wall during the lunar mornings. It may or may not be an impact crater. There is also a small nearly rimless craterlet "W" low on the west-northwest wall north of the west-rim bow-in which is about 2 miles across. It is considerably more difficult to observe than its size would indicate, as the Sun has to be at just the right angle to allow any shadow to fill even part of it to make it visible. (Credit: NASA)

the south are several rises collectively named the Montes Teneriffe. To the north lies the wide stretch of the Mare Frigoris. East of the crater, among the Montes Alpes, are several rilles collectively named the Rimae Plato.

Plato is about 3.84 billion years old, only slightly younger than the Mare Imbrium to the south. The rim is irregular with 2-km-tall jagged peaks that project prominent shadows across the crater floor when the Sun is at a low angle. Sections of the inner wall display signs of slumping, most notably a large triangular slide along the western side.

Plato has a reputation for being the site of transient lunar phenomena, including flashes of light, unusual color patterns, and areas of hazy visibility. These anomalies may be the result of seeing conditions, combined with the effects of different illumination angles of the Sun. The seventeenth-century astronomer Hevelius called this feature the 'Great Black Lake' because of the darkness of Plato's floor.

Plato's rim is circular, but from Earth it appears oval due to foreshortening.

According to old measurements reported in Thomas Gwyn Elger's 1895 book, *The Moon*, three peaks on the eastern rim rise 1.5, 1.8, and 2.1 km above the floor. On the western rim an obvious, large triangular massif is partially disconnected from the crater rim. This 15-km-long block, and another one farther north, result from giant landslides, where segments of the rim slid slightly inward, creating a scallop—a bite out of the circular rim. Variations in rim height and width are thus due to slumping, but the height differences on Plato's east rim must be of older, unknown origins.

Lack of a central peak is one Plato mystery with a simple solution. Compared to other craters of similar size, Plato should have a 2.2-km-high mountain rising from its floor. However, since Plato is filled with a 2.6-km layer of lava, the peak must be buried.

How many craterlets can you see on Plato's floor? Although some observers have reported dozens, only four craters are sufficiently large to be seen with a moderate telescope even when observing conditions are steady and the illumination angle is favorable. The largest of the four has a 2.2-km diameter.

For more than a century the floor of Plato has been the focus of intense quasi-scientific debate over suspected lunar changes. Three types of observations caused controversy: detection of small craters on Plato's floor, variation in floor darkness with changing Sun angles, and obscurations of the floor itself. Because the floor

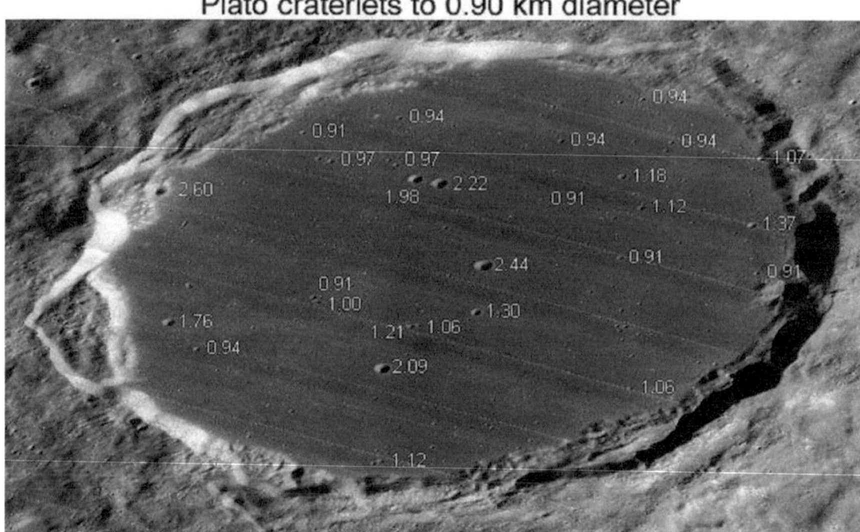

FIG. 4.2 The white annotations are craterlet diameters (in km) as measured on the lunar orbiter image with LTVT. Note that some features that look like they should be labeled, but aren't, are groupings of closely spaced smaller craterlets. For example, the prominent feature directly above the 1.76 km label on the left is two closely spaced craterlets with diameters of 0.79 and 0.82 km—together they are larger than some of the labeled single craterlets. The craterlets on the floor of Plato are one of the favorite targets for amateur astronomers testing the detection limit of their telescopes. (NASA)

possesses a few small impact craters near the limit of visibility with small telescopes, there have been unofficial contests to detect the largest number of craters (Fig. 4.2).

Harvard astronomy professor W. H. Pickering apparently won an unofficial contest in 1892 by announcing his mapping of 71 spots on Plato's floor. Comparing hand-drawn maps with high-resolution photographs obtained by the lunar orbiter spacecraft in 1967 demonstrates that the observers did detect the four largest craters, and some of the smaller ones, but their estimates of sizes, locations, and numbers were often seriously in error.

Craterlets were not the only source of controversial observations. According to Elger, who was an English lunar mapper and the first director of the Lunar Section of the British Astronomical

Association in 1895, "The gradual darkening of the floor of Plato as the sun's altitude increases from 20° till after full Moon may be regarded as an established fact, though no feasible hypothesis has been advanced to account for it." Actually, just the opposite is true, according to measurements of the floor's brightness by sensitive photometers mounted on large telescopes. Like the rest of the Moon, Plato's floor brightens until near full, when it rapidly gets much brighter, and then darkens after full Moon.

The notion that Plato's floor darkens in an absolute sense as the Sun rises over it, which Elger takes as an "an established fact," seems to have originated with British amateur W. R. Birt in about 1871; and was repeated not only by Elger (1895), but also by Neison (1876) and Goodacre (1931). A little before 1900, American astronomer W. H. Pickering reported similar areas of darkening in the craters Alphonsus, Riccioli and Atlas. Like Birt, Pickering assumed the darkening was evidence of physical change—in Pickering's mind, most probably a monthly ebb and flow of lunar vegetation. As photometric measurements in the mid-twentieth century demonstrated, the darkening is simply an illusion: all surface areas brighten with the approach of Full Moon, but mare areas—like the floor of Plato—brighten less strongly, making them appear darker by contrast.

The third of Plato's controversies concerns reports that the dark floor is occasionally obscured by mists or clouds. Most of the observations were made during the last century; Walter Goodacre's 1931 book, also called *The Moon*, mentions that there are "a number of well authenticated cases." Descriptions include a fog that cleared as the Sun rose, and a "curious luminous milky kind of light." Another nineteenth-century observer found that the floor was covered by myriad points of light, "as if reflected from flocculent clouds lying near the surface."

The area to the northeast of Plato appears to be empty, and to an untrained eye it would possibly appear to be uninteresting. However, nothing could be farther from the truth. Radar studies of this region, along with spectroscopic data from the US Clementine mission indicate that this region may be covered with pyroclastic materials. As an added bonus there is a large sinuous rille, possibly related to the pyroclastic deposit, just south of the site, close enough that astronauts could explore it using a simple rover.

Pyroclastic deposits consist of small, glassy volcanic beads that cover the surrounding terrain. They were produced by explosive fire-fountaining eruptions billions of years ago when the Moon was more geologically active. When magma cools rapidly (quenched), atoms do not have the time to form orderly structures (minerals) so a glass results.

Pyroclastic glasses are actually the most primitive (that is, unmodified by later geologic processes) materials in the lunar sample collection. The study of the pyroclastic materials in the Apollo collection has given geochemists valuable insights into the composition and evolution of the lunar interior. However, there are comparatively few pyroclastic materials in the current collection, and the Apollo sites themselves are not geologically representative of all lunar surfaces. Human lunar exploration is needed to complete our understanding of the formation and history of the Moon.

Pyroclastic deposits are some of the most potentially valuable resources to support human habitation on the Moon. Processing them can provide relatively easy access to oxygen and water to future lunar explorers. Human exploration of this location will enable access to these important resources, as well as provide key insights into the nature of these pyroclastic materials and the possible source of the volcanic eruptions that produced them.

Reiner Gamma

The Moon has several bizarre swirl patterns on the surface that look as if visiting aliens took spray paint and left graffiti—"Zord was here." Reiner Gamma is one of the strangest features on the Moon. It is the best-known example of a lunar swirl that has baffled planetary geologists for decades. This peculiar pattern looks a little like crater rays, but it's curvy and lacks a source crater. In fact, there's nothing obviously geological about it. It remains one of the big unanswered questions in lunar science (Fig. 4.3).

The mystery only deepened when orbiting Apollo spacecraft (and later *Lunar Prospector*) found strong magnetic fields directly over the swirls—surely an important clue. One theory is that the patchy, localized magnetic field is shunting solar-wind plasma

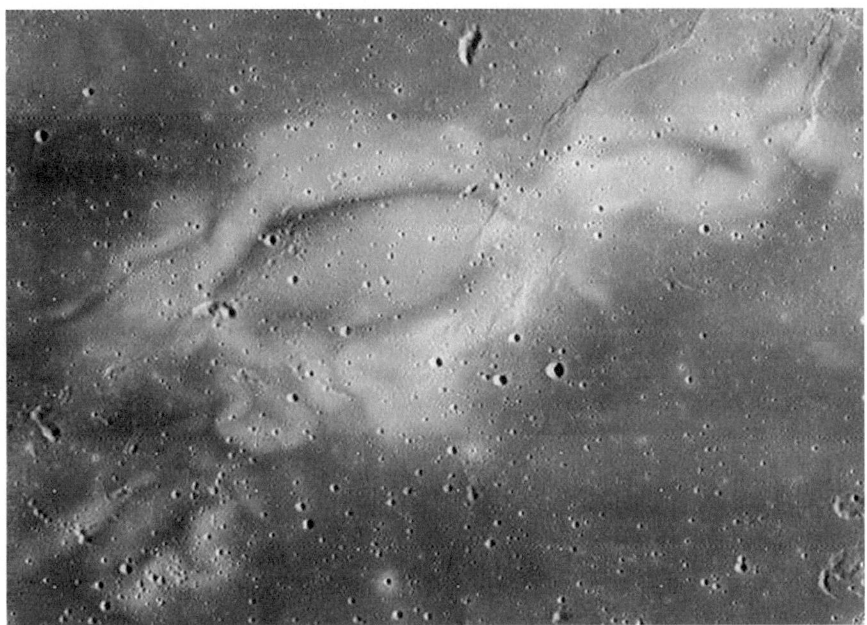

Fig. 4.3 The mysterious tadpole swirl Reiner Gamma is about 20 miles (35 km) across. This view was captured by a lunar orbiter in 1967. (Credit: NASA)

away from the swirls' bright areas and funneling it to the dark lanes that snake through the features.

First identified by early astronomers during the Renaissance, the Reiner Gamma formation has been a subject of intense scientific study for centuries. Reiner Gamma is one of the most distinctive natural features of the Moon. This striking, tadpole-shaped swirl has puzzled planetary geologists for decades. It is located on the Oceanus Procellarum, to the west of the crater Reiner. The center of the formation is located at selenographic coordinates WikiMiniAtlas 7°30′N 59°00′W/7.5°N 59.0°W/7.5; –59.0. It has an overall dimension of about 70 km. The feature has a higher albedo than the relatively dark mare surface, with a diffuse appearance and a distinctive swirling, concentric oval shape. Related albedo features continue across the surface to the east and southwest, forming loop-like patterns over the mare.

Three volumes filled with studies of Reiner Gamma have been amassed since 1979 in an effort to understand this strange enigma. The theories of Reiner Gamma's formation abound, but none have been tested, or confirmed through observation. Among the pre-

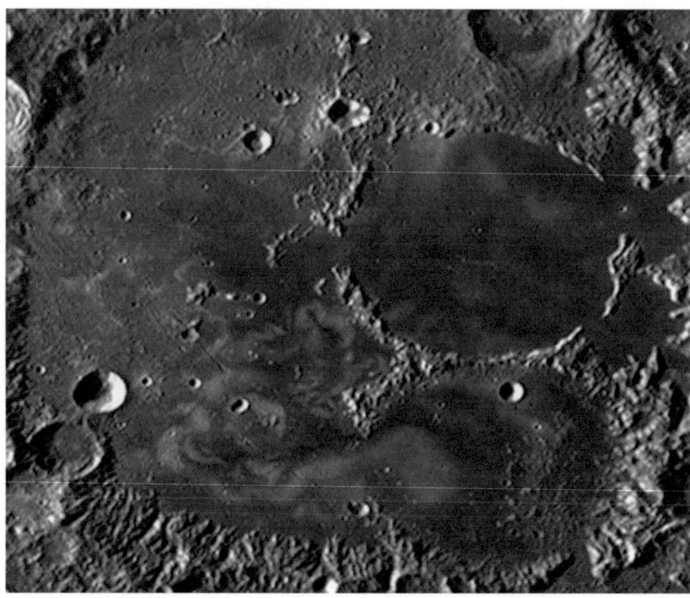

FIG. 4.4 Light swirls at the SW sector of Mare Ingenii (Credit: NASA)

vailing theories are: (a) high-velocity comet impact; (b) seismic activity; (c) latent magnetic fields from large, rayed craters nearby; (d) magnetic storms from the Sun; (e) lunar volcanism (Fig. 4.4).

A high velocity comet impact would not have left a mere surface deposit of powdered debris. There would have been a sizeable impact crater, as well as large boulders and other rock deposits from the blast effect. Those features are missing. Seismic activity/lunar volcanism could explain the dust-like deposits that appear to have "settled out" from the cloud of an eruption event. However, there is no volcanic vent, nor large crack out of which the material could have been ejected. Further, lunar quakes of sufficient magnitude have never been detected by instrument packages left behind by various Apollo missions to the Moon.

Recent magnetic data suggest a link to impact-related magnetism because the swirls in Mare Ingeni are associated with relatively strong magnetic anomalies antipodal to some of the young impact basins such as Imbrium and Crisium.

Some scientists believe Reiner Gamma is a surface deposit, similar to the powdered rock rays extending from lunar craters like Tycho. Rayed features from craters are not the result of explosive ejection, but are the remains of electric discharges moving across the lunar surface. According to Ralph Juergens, one of a few scientists who, during his lifetime, acknowledged and explained the electric cosmos:

"It's all-but-impossible to reconcile with ejection origins. Enormous velocities of ejection must be postulated to explain the lengths of the rays (of Tycho), yet the energetic processes responsible for such velocities must be imagined to be focused very precisely to account for the ribbon-thin appearance of the rays".

Latent magnetic fields are part of the formation (as well as those in Mare Marginis and Mare Ingenii) and were measured by the Magnetometer/Electron Reflectometer (MAG/ER) experiment onboard the Lunar Prospector spacecraft in the 1990s. Measurements of the remnant magnetism in the lunar crust were taken by observing the magnetic reflection of electrons from the Moon's surface. The differences are small, but the evidence points to a variable strength magnetic field imprinted on the Moon. Since magnetism is a result of electrical currents, then the swirling pattern of the formation may be the result of an external electrical influence.

Swirls, arcs and curlicues are found on every rocky body in the solar system, as well as on most of the ice-bodies. An explanation for the swirling path of Reiner Gamma must include the features of varying scale that are found on the other Moons. According to electrical theorists, a common explanation for each of the geo-morphological features that we have illustrated is that the planets and Moons exist within an electric universe. Each solar system object is normally insulated within its individual charge sheath. However, when those sheaths touch, electric charge is exchanged.

Therefore, we find that the Moon, Io, which orbits within Jupiter's plasma sheath, is undergoing continuous electrical bombardment from its parent planet.

Reiner Gamma is one more example of surface scarring in an electrically active Solar System. The braided appearance and the deposit's lack of depth suggest rotating Birkeland currents (set of currents which flow along geomagnetic field lines connecting the

Earth's magnetosphere to the Earth's high latitude ionosphere). These are draped over the surface, creating a filamentary and pleated look to the light-colored field. The shape of the deposit is suggestive of the auroral curtains that can be seen in the skies of Earth's polar regions. Reiner Gamma may be the result of similar activity by energetic streamers from the sun or another charged body.

A concept that comes up frequently in lunar science is "space weathering." This term is used to described a suite of natural processes (including micrometeoroid impacts and exposure to the solar wind) that can alter the spectral properties of lunar surface materials. Since the reflectance of the lunar surface within the Reiner swirl is so different from the surrounding mare, some process may have altered the space weathering susceptibility of the swirl materials.

The central feature of Reiner Gamma does bear a resemblance to the dipolar formation created by iron filings on a surface with a bar magnet on the underside. Low-orbiting spacecraft have observed a relatively strong magnetic field associated with each of these albedo markings. Some have speculated that this magnetic field and the patterns were created by cometary impacts. However the true cause remains uncertain.

There are several theories to account for the presence of the Reiner Gamma swirl. Results from previous lunar missions (including Lunar Prospector) have indicated that the swirl region has an elevated magnetic field, so it's possible that an event hundreds of millions of years ago modified the magnetic properties of the surface materials, deflecting the solar wind and changing how the reflectance is modified by space weathering. Some investigators have proposed that the coma of a comet—streaking in just above the surface—interacted with the lunar surface, changing the surface properties to the degree where the Reiner Gamma swirl could persist for millions of years.

V. V. Shevchenko of Sternberg State Astronomical Institute in Moscow, long a proponent of the "recent cometary influence" theory concerning the origin of Reiner Gamma, has pointed to a faint structure of increased crater count that overlaps a seemingly uninterrupted background noise of craters that is apparently coincident with some of the eastern side of Reiner Gamma's huge eye.

This is definitely seen in earlier laser altimetry of the area and again in the LOLA image.

However, many researchers feel that the answer is not the influence of comets or cometary tails, as some have long suggested. Rather, there now appears to be evidence of a link with the deposition of peculiar kinds of the nanophase iron found everywhere on the Moon, perhaps made peculiar by the influence of the magnetic anomalies themselves, and daily transport of newly gardened alternately ionized submicron-sized lunar dust. Still, the nature of the Moon's local magnetism is clearly not the same in every situation. Perhaps something a little different is happening at Reiner Gamma.

Reiner Gamma is not associated with any particular irregularities in the surface, and so the cause was a mystery until similar features were discovered in Mare Ingenii and Mare Marginis on photographs taken by orbiting spacecraft. The feature on Mare Ingenii is located at the lunar opposite point from the center of Mare Imbrium. Likewise the feature on Mare Marginis is opposite the midpoint of Mare Orientale. Thus it is believed that the feature resulted from seismic energies generated by the impacts that created these maria. Unfortunately there is no such lunar mare formation located precisely on the opposite surface of the Moon, although the large crater Tsiolkovskiy lies within one crater diameter (Fig. 4.5).

Recent observations being brought to bear on Reiner Gamma include scans by a magnetometer aboard Lunar Prospector and topographic mapping by a laser altimeter aboard Lunar Reconnaissance Orbiter. Fortunately, recent missions to the Moon—particularly NASA's Lunar Reconnaissance Orbiter (LRO), India's Chandrayaan 1, and Japan's Kaguya—have amassed a formidable arsenal of new observations, and a score of specialists gathered at the recent Lunar and Planetary Science Conference to plan their analytical attack.

So far, the new data are saying a lot about what lunar swirls *aren't*. For example, tedious crater-counting by Kramer and others is used to show that the Mare Ingenii and other swirls are not freshly exposed surfaces. LRO's laser altimeter shows that they're not lumpy or elevated; radar scans haven't turned up any differences in surface roughness, and infrared maps made during the long lunar night reveal no thermal anomalies.

Fig. 4.5 Mare Marginis swirls (Credit: NASA)

One promising idea, put forward at last year's meeting by Ian Garrick-Bethell and two colleagues at Brown University, avoids the pitfalls of selective space weathering and instead envisions a way to keep the swirls bright with a continually replenished topping of very fine dust. A series of television images relayed to Earth in January 1968 by the lunar lander *Surveyor 7* show a line of light along its western horizon after sunset—the result of fine dust particles suspended above the surface. *Surveyors 5* and *6* witnessed the same phenomenon.

Their view is based on the unexpected revelation that microscopic particles become statically charged and levitated above the lunar surface, particularly at times near local sunrise and sunset. (No one knows why this happens, but it does.)

Once lofted, these ultrafine particles can be moved around by electric fields created within the localized magnetic bubbles, lateral shuffling that removes dust in some places and deposits it in others. The dusty topcoat would have to be at least a foot thick, based on calculations published in *Icarus*. But they can't be thicker than about 30 ft (10 m), notes Garrick-Bethell, because the swirls' bright lanes show no vertical relief in low-Sun-angle images.

In time, the new infusion of composition and reflectivity measurements might finally determine the true nature of these delicate, enigmatic features, though Garrick-Bethell cautions, "Ultimately, we really need some measurements of the plasma and dust environment near the surface."

How and when the magnetic anomalies arose is yet another mystery. One idea, proposed at the meeting by Mark Wieczorek (Institute of Earth Physics of Paris), posits that they're linked to blobs of molten debris tossed out during the titanic collision that formed the far side South Pole-Aitken basin.

Reiner Gamma's magnetic field strength is approximately 15 nT, measured from an altitude of 28 km. This is one of the strongest localized magnetic anomalies on the Moon. The surface field strength of this feature is sufficient to form a mini-magnetosphere that spans 360 km at the surface, forming a 300 km thick region of enhanced plasma where the solar wind flows around the field. As the particles in the solar wind are known to darken the lunar surface, the magnetic field at this site may account for the survival of this albedo feature.

In early lunar maps by Francesco Maria Grimaldi, this feature was incorrectly identified as a crater. His colleague Giovanni Riccioli then named it *Galilaeus*, after Galileo Galilei. The name was later transferred northwest to the current crater Galilaei.

Much of the surface experienced significant tectonic activity, resulting in flexing and faulting of the surface. In fact, several wrinkle ridges are prominent throughout the Reiner Gamma region. However, further study is needed to determine the relative ages of the faulting and the albedo anomaly at Reiner Gamma. Specifically, did the compression stresses associated with the formation of wrinkle ridges deform the albedo anomaly, or is the albedo anomaly draped over pre-existing faults and folds? These questions (and more) about Reiner Gamma still remain unanswered.

The analysis of NASA's Clementine imaging data showed that the optical and spectroscopic properties of the local regolith surface layer are close to those of immature mare crater-like soils. This is consistent with the properties of a shallow subsurface mare soil layer.

Future high resolution study of both the composition and structure of Reiner Gamma is yet needed. However, the swirls are likely formed through a surface process that does not deposit or remove large volumes of additional materials.

The Reiner Gamma formation would be a fascinating site for future human exploration. They wouldn't find a monolith, but if astronauts brought along a magnetic compass the needle would wildly swing back and forth. And, the localized field would helpfully shield them from the solar wind.

The Aristarchus Plateau

The Aristarchus plateau is one of the most geologically diverse places on the Moon; it has a mysterious raised flat plateau, a giant rille carved by enormous outpourings of lava, fields of explosive volcanic ash, and all this is surrounded by massive flood basalts. Scientists think the crater was created relatively recently, geologically speaking, when a comet or an asteroid smashed into the Moon, gouging out a hole in its surface (Fig. 4.6).

The Aristarchus plateau is an elevated rocky rise located in the midst of the Oceanus Procellarum, a large expanse of lunar mare. This is a tilted crustal block, about 200 km across, that rises to a maximum elevation of 2 km above the mare in the southeastern section.

The Aristarchus plateau has fascinated lunar observers since before the Space Age. Its odd shape and low and high albedo extremes immediately draw your attention. Superimposed on the plateau is a spectacular channel (or rille), and the very young Aristarchus Crater. Aristarchus Crater is the largest impact crater on the plateau and is one of the highest reflectance features on the Moon.

The plateau itself is surrounded by the lava flows of Oceanus Procellarum, and the whole region has a high concentration of sinuous rilles. The largest of these is Vallis Schröteri, which is also the largest sinuous rille on the Moon. Finally, the plateau is almost completely covered by one of the largest lunar regional pyroclastic deposits. Large pyroclastic deposits are a potential resource for useful elements such as hydrogen, oxygen, iron and titanium.

The head of Vallis Schröteri, a feature also known as the "Cobra Head," consists of a deep pit, which is of great interest to

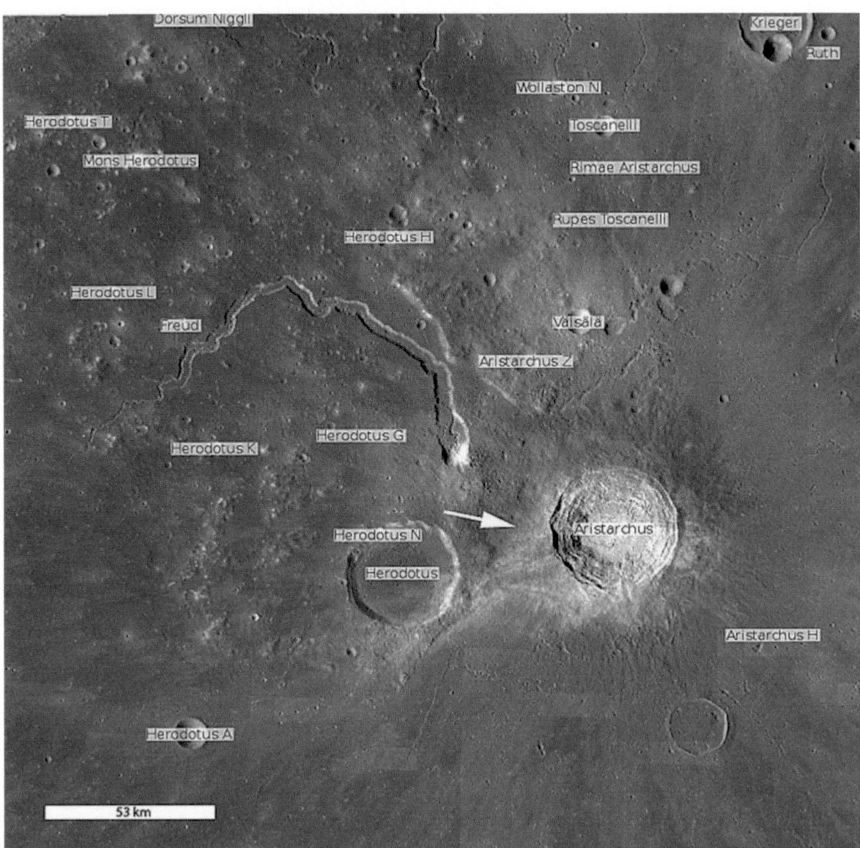

Fig. 4.6 LRO WAC regional view of Aristarchus plateau and crater. Aristarchus, named after the Greek astronomer Aristarchus of Samos, is a prominent lunar impact crater that lies in the northwest part of the Moon's near side. It is considered the brightest of the large formations on the lunar surface, with an albedo nearly double that of most lunar features. The region is bright enough to be visible to the naked eye and is dazzling in a large telescope. It is also readily identified when most of the lunar surface is illuminated by earthshine. (Credit: NASA)

scientists. The Cobra Head is thought to be the source vent of a tremendous outflowing of lava that flowed across the plateau and formed the rille. Exposed in this vent are lava flows, pyroclastic material, and small bits of white rocks. The white rocks are pieces of the underlying crust most likely composed of anorthositic (highland) rocks. Much of this material was excavated by the impact that formed Aristarchus Crater and was brought to the surface from great depths. These white rocks may be of a unique crustal composition representing late-stage subterranean magmatic activity. Rocks that form last from a magma body often have rare compositions enriched in incompatible elements. The elements are labeled incompatible because they do not easily combine with other elements and thus concentrate in the last remaining melt. Therefore, the last rocks to freeze out of a magma have high levels of these incompatible elements. Sampling such rocks will provide insights into lunar magmatic evolution and the bulk composition of the mantle (Fig. 4.7).

The slopes along the eastern wall of the Cobra Head are covered with boulders and debris that slid down the walls. The patterns on

FIG. 4.7 The west wall of the Moon's Aristarchus Crater seen obliquely by NASA's LRO from an altitude of only 16 miles (26 km), which is about two times lower than normal, due to LRO's current elliptical orbit. The spacecraft was slewed to the west for an oblique, or "sideways," look at the crater, instead of looking straight down as LRO normally does, to provide this unique perspective on Aristarchus. For a sense of scale that altitude is only a little over twice as high as commercial jets fly above Earth. This crater is only one-tenth the size of Earth's Grand Canyon, but the views from up above are similarly spectacular. (Credit: NASA)

the slopes are evidence for downslope movement (landslides and/ or creep). "Flow" is from right to left, with boulders accumulating in local "bars." The slope is scoured by the movement of boulders. The rocks exposed in the rille exhibit large brightness contrasts, some being very dark and others very bright.

What is the origin of the bright material? Some of the brightness is a function of steep boulder faces oriented towards the Sun. Even dark rocks such as basalt can exhibit bright Sun-facing facets. Most of the local material here is basaltic, having originated in the huge Cobra Head vent at the head of the 140-km long Vallis Schröteri. To the east is the Aristarchus Crater, which has some of the brightest ejecta of any crater on the Moon. Some of that bright ejecta material is among the boulders that have moved down this slope as the steep rim gradually collapsed over time.

The plateau has been the site of many reported transient lunar phenomena (LTP), with nearly 700 to-date—the highest recorded for any lunar feature. Such events include temporary obscurations and colorations of the surface, and catalogs of these show that more than one-third of the most reliable spottings come from this locale. No wonder planners for future missions put this plateau high on its list of targets for human exploration.

In 1971, when *Apollo 15* passed 110 km above the Aristarchus plateau, a significant rise in alpha particles was detected. These particles are believed to be caused by the decay of radon-222, a radioactive gas with a half-life of only 3.8 days. The Lunar Prospector mission later confirmed radon-222 emissions from this crater. These observations could be explained by either the slow and visually imperceptible diffusion of gas to the surface, or by discrete explosive events (Fig. 4.8).

On November 2, 1958, the Russian astronomer Nikolai A. Kozyrev observed an apparent half-hour "eruption" that took place on the central peak of Alphonsus Crater using a 48-in. (122-cm) reflector telescope equipped with a spectrometer. During this time, the spectra showed evidence for bright gaseous emission bands due to the molecules C_2 and C_3. While exposing his second spectrogram, he noticed "a marked increase in the brightness of the central region and an unusual white color." Then, "all of a sudden the brightness started to decrease," and the resulting spectrum was normal.

FIG. 4.8 This is a view of the Aristarchus and Herodotus craters taken from orbit during the *Apollo 15* mission. The view is toward the south. Aristarchus Crater is near the center of the image, and the flooded Herodotus is to the right. About a quarter of the way up from the bottom edge is Väisälä Crater, a crisp-edged, bowl-shaped feature with a smaller craterlet near its right edge. The wide cleft below Herodotus is the Vallis Schröter. A rille system in the lower left is part of the Rimae Aristarchus. Note how these rilles appear to begin at the edge of craters or depressions. Small secondary impacts from the Aristarchus impact are visible as streak-like marks and craterlets in the lunar mare to the south of the rilles. The fault line at the bottom is the Rupes Toscanelli. The lunar mare in the top part of the image belongs to the Oceanus Procellarum. Some wrinkle-ridges are visible. (Credit: NASA)

During the *Apollo 11* mission in 1969, Houston radioed to *Apollo 11*: "We've got an observation you can make if you have some time up there. There's been some lunar transient events reported in the vicinity of Aristarchus." Astronomers in Bochum, West Germany, had observed a bright glow on the lunar surface— the same sort of eerie luminescence that has intrigued Moon watchers for centuries. The report was passed on to Houston and then to the astronauts. Almost immediately, Armstrong reported back: "Hey, Houston, I'm looking north up toward Aristarchus now, and there's an area that is considerably more illuminated than the surrounding area. It seems to have a slight amount of fluorescence."

In another case occurring near the Cobra Head feature on the Aristarchus plateau, there is good quantitative confirmation on Clementine images for the LTP observed from ground. On April 23, 1994, ground-based observers reported a "possible obscuration over Cobra Head" between 03:50 and 04:30 UT. An analysis of the Clementine images taken on 1994, March 3 and April 27, show a measurable difference in color. The images obtained on April 23 exhibit a 15 % increase in color ratio (0.4–1.0 μm) when compared to the images acquired on March 3. The Cobra Head, which is a collapsed lava tube, is an area where many past observations of LTPs have been reported. These include a reddish-orange brightening lasting at least 25 min.

More recently, Japan's Selene (Kaguya) lunar mission detected radioactive alpha particles from Radon outgassing while flying over Aristarchus.

Aristarchus is considered the brightest of the large formations on the lunar surface, with an albedo nearly double that of most lunar features. The feature is bright enough to be visible to the naked eye, and is dazzling in a large telescope. It is also readily identified when most of the lunar surface is illuminated by earthshine (Fig. 4.9).

Aristarchus is bright because it is a young formation, approximately 450 million years old, and the solar wind has not yet had time to darken the excavated material by the process of space weathering. The impact occurred following the creation of the ray crater Copernicus, but before the appearance of Tycho. Due to its prominent rays, Aristarchus is mapped as part of the Copernican system.

The brightest feature of the Aristarchus Crater is the steep central peak. Sections of the interior floor appear relatively level, but lunar orbiter photographs reveal the surface is covered in many small hills, streaky gouges, and some minor fractures. The crater has a terraced outer wall, roughly or polygonal in shape, and covered in a bright blanket of ejecta. These spread out into bright rays to the south and southeast, suggesting that Aristarchus was most likely formed by an oblique impact from the northeast, and their

Fig. **4.9** *Apollo 15* photo of Aristarchus Crater (the brighter of the two large craters near the top of the image). About 40 km across, Aristarchus is one of the brightest features on the Moon and the sight of many transient lunar phenomenon. Beside it is Herodotus Crater. Closer to the camera (which was looking south) is the superb sinuous rille Vallis Schroteri. (Credit: NASA)

composition includes material from both the Aristarchus plateau and the lunar mare.

In November 2011, the Lunar Reconnaissance Orbiter passed over the crater, which spans almost 25 miles (40 km) and sinks more than 2 miles (3.5 km) deep. At that time, Professor Robert W. Wood used ultraviolet photography to take images of the crater area. He discovered the plateau had an anomalous appearance in the ultraviolet, and an area to the north appeared to give indications of a sulfur deposit. This colorful area is sometimes referred to as "Wood's Spot," an alternate name for the Aristarchus Plateau.

Spectra taken of this crater during the Clementine mission were used to perform mineral mapping. The data indicated that the central peak is a type of rock called anorthosite, which is a slow-cooling form of igneous rock composed of plagioclase feldspar.

By contrast the outer wall is troctolite, a rock composed of equal parts plagioclase and olivine.

The Aristarchus region was part of a Hubble Space Telescope study in 2005 that was investigating the presence of oxygen-rich glassy soils in the form of the mineral ilmenite. Baseline measurements were made of the *Apollo 15* and *Apollo 17* landing sites, where the chemistry is known, and these were compared to Aristarchus.

The Hubble Advanced Camera for Surveys was used to photograph the crater in visual and ultraviolet light. The crater was determined to have especially rich concentrations of ilmenite, a titanium oxide mineral that could potentially be used in the future by a lunar settlement for extracting oxygen.

Many of the enigmatic features emanating from the rim of Aristarchus are suggested to be volcanic features. This debate raged throughout the 1960s and 1970s, but most modern interpretations of flow features associated with Copernican-aged impact craters are attributed to impact-generated melt rock. Aristarchus contains interesting examples of sinuous channels, which appear to have fed large flow fronts extending away from the crater. The higher resolution of LROC data allows these features to be studied in greater detail. Clear evidence of melt flow occurring subsequent to the formation of the crater rim is observed, and superimposed craters have also been deformed by the flow. Flow tongues are observed to cross bright ray material, indicating the melt was molten for some time after the main excavation stage.

On November 10, 2011, the LRO made a low-altitude pass 42 miles to the east of Aristarchus Crater and took an amazing panoramic image of the western wall of the crater. The LRO was flying at just over 16 miles above the lunar surface at the time, which is about two times lower than normal, due to the LRO's current elliptical orbit. As a result, the image has incredible detail. The spectacular new view from the LRO is almost as good as being there. In fact, the video lets you "rappel" down and take a closer look at the western side of the crater walls.

The edges of the wall appear scalloped, almost like the crater was strip-mined. Since the crater is relatively young, Aristarchus is one of the brightest regions on the Moon. These bright rocks may be anorthositic, like the highlands, or they may be a more silicic rock like granite—or both. "Although granites have been found in Apollo rock samples, the formation of granite on the Moon is not well understood at this time—another reason why we need to get samples from this region," said LRO Camera Principal Investigator Mark Robinson.

"Diverse materials such as dark, multilayered mare basalts in the walls, bright crustal rocks in the central peak, impact melt, and even regional pyroclastic materials blanketing the crater are brought to the floor and accumulated through mass wasting (a geomorphic process by which soil, sand, regolith, and rock move downslope typically as a mass, largely under the force of gravity) creating a bountiful trove of geologic materials," Robinson said.

Looking up and over to the east, on the steep walls of the crater, you can observe the streaks marking ejecta leaving the crater in the LRO images. The shades of gray in the streaks likely trace back to the diverse stratigraphy seen in the central peak. Imagine a future lunar explorer traversing around the base of the 300-meter-tall central peak while collecting samples.

Crater central peaks are formed as the Moon's crust rebounds after the tremendous stress of an impact is released. The energies of impacts are so high that rocks no longer behave as brittle solids, but rather as deformable plastic. It's more accurate to think of the crust behaving like a fluid; as the crater forms, the bottom of the crater is first pressed down, then it rebounds. For craters above 20 km in diameter, the rebound is so strong that material from the depth is actually brought up and forms a central peak. In the case of Aristarchus Crater, the central peak contains rocks with three very different albedos. As these materials erode out of the central peak they slide downslope, creating contrasting stripes. The highest albedo material reflects about four times as much light as the lowest albedo rocks.

What causes these extreme differences in albedo? From other datasets (Earth-based telescopes, Clementine, Chandrayaan, Kaguya) it has long been known that the interior of Aristarchus crater is compositionally heterogeneous. The high albedo material is most likely a common lunar rock type, anorthosite. The bulk of the lunar crust is made of anorthosite; it forms the bright material that you see when you look at the Moon (the dark areas are basaltic). On the other hand, perhaps we are seeing a more silicic rock akin to granite? Such silica-rich rocks are known to form on the Moon; we just do not know much about their origin and locations.

What is the dark material exposed in the Aristarchus central peak? Scientists are not sure; however we know the adjacent Aristarchus plateau is blanketed in dark pyroclastic deposits. "Pyroclastic" derives from the Greek words for fire and broken, as in the small, hot broken rocks erupted in an explosive volcanic event. It is likely that the dark material is related to the nearby pyroclastics; perhaps the impact excavated a now solidified dike that once carried volcanic material to the surface.

To truly understand the geology of a location scientists typically examine the subsurface as well as the surface. On Earth geologists can drill holes, go into mines, and inspect road cuts. Planetary scientists must rely on impact craters to expose subsurface rocks—nature provides us with natural drill holes! The WAC (Wide Area Camera) color image dramatically shows regional differences, most likely due to changes in rock type. These colors are different than what your eye would see.

First, the WAC is sensitive to ultraviolet wavelengths (300–400 nm) where some minerals (ilmenite for one) are highly absorptive, thus providing a marker of their presence, like a fingerprint. Secondly, images that contain subtle color differences can be stretched to bring out small variations. Mapping of color units that correspond to rock boundaries will guide future planners in terms of identifying the diversity of rock types. How do these broad color differences relate back to the boulders seen lying on the crater floor? Subtle color differences may not be readily apparent to the naked eye. Future explorers would want to carry portable color analyzers (spectrometers) to help sort through the boulders they will encounter on the surface.

There are at least three different distinct shades of grey seen on the steep slopes, perhaps more? Though climbing to the top and walking the ridge would afford a magnificent view, simply collecting rocks along the base would suffice from a geologic viewpoint. The next logical step is to collect samples along the east and southeast wall of the crater. Do the distinct rays seen trace back to the layers seen in the central peak? What about the enigmatic reddish brown material just on the southeast flank? The geology of the Moon is rich and complicated!

From LROC WAC images you can see that the gray streaks show up as distinct color anomalies, color due to variations in rock type. The area has long been known to be among the reddest spots on the Moon—meaning its reflectance strongly increases from short to long wavelengths. In the WAC color image, you can see the distinct red-hued region, which is largely blanketed by the glass-rich products of explosive volcanic eruptions. This area is surrounded by bluer terrain, which formed when titanium-rich (at least it is thought that titanium is in these rocks) lava flowed across the surface and flooded the area, forming a portion of Oceanus Procellarum. In both of the WAC context images, you can see Vallis Schröteri, a canyon-like feature, known as a sinuous rille, through which the lava once flowed.

At the intersection of this amazingly diverse region, the Aristarchus impact event gives us a three-dimensional look into the plateau. The central peak of Aristarchus, which contains uplifted material from great depths, is thought to in part contain the anorthositic material that makes up the lunar highlands, and portions of the southeastern wall and rim, which are seen in today's oblique WAC view of the crater, are thought to be rich in olivine.

The bounty of geologic processes that came together to produce this complex region makes it a high-priority target for future exploration. The geologic interest is not the end of the story. Pyroclastic deposits there may contain valuable resources to be mined by future explorers.

Shackleton Crater

As much as 22 % of the surface material in a crater located on the Moon's south pole may be made up of ice, according to data from the LRO spacecraft (Fig. 4.10).

The team of NASA and university scientists using laser light from LRO's laser altimeter examined the floor of Shackleton Crater. They found that the crater's floor is brighter than those of other nearby craters, which is consistent with the presence of small amounts of ice. This information will help researchers understand crater formation and study other uncharted areas of the Moon. The findings are published in the journal *Nature*.

FIG. 4.10 This is an elevation map of Shackleton Crater was made using LRO Lunar Orbiter Laser Altimeter data. The false colors indicate height, with blue lowest and *red/white* highest. (Credit: NASA)

"The brightness measurements have been puzzling us since two summers ago," said Gregory Neumann of NASA's Goddard Space Flight Center in Greenbelt, MD, a co-author on the paper. "While the distribution of brightness was not exactly what we had expected, practically every measurement related to ice and other volatile compounds on the Moon is surprising, given the cosmically cold temperatures inside its polar craters."

The spacecraft mapped Shackleton Crater with unprecedented detail, using a laser to illuminate the crater's interior and measure its albedo, or natural reflectance. The laser light measures to a depth comparable to its wavelength, or about a micron. That represents a millionth of a meter, or less than one ten-thousandth of an inch. The team also used the instrument to map the relief of the crater's terrain based on the time it took for laser light to bounce back from the Moon's surface. The longer it took, the lower the terrain's elevation.

In addition to the possible evidence of ice, the group's map of Shackleton revealed a remarkably preserved crater that has remained relatively unscathed since its formation more than 3 billion years ago. The crater's floor is itself pocked with several small craters, which may have formed as part of the collision that created Shackleton.

The crater, named after the Antarctic explorer Ernest Shackleton, is 2 miles deep and more than 12 miles wide. Like several craters at the Moon's south pole, the small tilt of the lunar spin axis means Shackleton Crater's interior is permanently dark and therefore extremely cold.

"The crater's interior is extremely rugged," said Maria Zuber, the team's lead investigator from the Massachusetts Institute of Technology. "It would not be easy to crawl around in there."

Although the crater's floor was relatively bright, Zuber and her colleagues observed that its walls were even brighter. The finding was at first puzzling. Scientists had thought that if ice were anywhere in a crater, it would be on the floor, where no direct sunlight penetrates.

The upper walls of Shackleton crater are occasionally illuminated, which could evaporate any ice that accumulates. A theory offered by the team to explain the puzzle is that "Moonquakes"—seismic shaking brought on by meteorite impacts or gravitational tides from Earth—may have caused Shackleton's walls to slough off older, darker soil, revealing newer, brighter soil underneath. Zuber's team's ultra-high-resolution map provides strong evidence for ice on both the crater's floor and walls (Fig. 4.11).

"There may be multiple explanations for the observed brightness throughout the crater," said Zuber. "For example, newer material may be exposed along its walls, while ice may be mixed in with its floor."

The initial primary objective of LRO was to conduct investigations that prepare for future lunar exploration. Launched in June 2009, LRO completed its primary exploration mission and is now in its main science mission. LRO was built and is managed by Goddard. This research was supported by NASA's Human Exploration and Operations Mission Directorate and Science Mission Directorate at the agency's headquarters in Washington.

Though unremarkable in appearance compared to the roughly 4000 craters on the Moon in its size range, the 20-km-diameter crater Shackleton has been the source of relentless scientific

FIG. 4.11 The south pole of the Moon is at 10 o'clock on the rim of Shackleton Crater

controversy for the past 20 years. Shackleton is located at the precise location of the geographic south pole of the Moon. Its location makes observation by Earth-based telescopes difficult, and it was not well photographed by the Lunar Orbiter series (our principal source of lunar images) of the 1960s. That all changed in 1994 with the flight of the joint DOD-NASA mission to the Moon, Clementine.

Clementine carried cameras that globally imaged the Moon in 11 visible and near-infrared wavelengths. In addition, it mapped the surface and lighting of the poles of the Moon at uniform resolution over the course of almost 3 lunar days (74 Earth days). When the science team first saw the south polar mosaic, the extent of darkness in the map was striking. Because the Moon's spin axis is close to perpendicular to the ecliptic plane, the Sun is always at the horizon at the lunar poles. Instead of rising and setting, the Sun circles around the poles at or near the horizon. Because of this grazing incidence, an area in a topographic depression may be in permanent shadow. And so it appeared for Shackleton Crater in the Clementine data, setting off bells in the heads of the science team.

A key controversy of the post-Apollo era was whether the lunar poles might contain water or not. Although the Apollo samples had been studied and found to be "bone-dry," we had not been to the poles on any Apollo mission. We knew that any shadowed areas had to be extremely cold as well as permanently dark. As water-bearing debris in the form of asteroids and comets constantly strike the Moon, it was thought that some of that water might get into a polar "cold trap" and would be kept there (essentially) forever; billions of years of impacting cosmic "debris" can add up.

Clementine was not configured to measure the presence of water, but a cleverly improvised experiment used the spacecraft's data transmitter to beam radio waves into the dark regions near the poles and listen to their reflected echoes on the enormous (70-m) dish antenna of NASA's Deep Space Network. Interestingly, the reflections indicated an enhancement of "same sense" polarization within the (very large) resolution cell that contained Shackleton crater. A collection of data from a nearby sunlit area (taken as an experimental control) did not show this peak. The Clementine team interpreted the RF peak as evidence for the presence of a few percent of water-ice within the dark, cold interior of Shackleton Crater. The media quickly spread the startling news about water on our "bone-dry" Moon.

Such a controversial conclusion did not go unchallenged. Some in the radar community argued that abundant wavelength-sized rocks on the surface were the source of the enhanced reflection. Since the lunar surface is indeed rocky, this interpretation could not be ruled out. Then a few years later, the Lunar Prospector mission found an enhancement of hydrogen concentration at both poles of the Moon; as hydrogen is a major constituent of water, the idea ice exists in the dark areas gained credence and has led to a decade-long scientific search (using a variety of techniques) for lunar polar ice. Though many areas near the poles were studied in detail, attention continued to be drawn back to Shackleton and the area near the south pole (Fig. 4.12).

Many different, new sensors have flown to the Moon in the last few years, including radar, ultraviolet (UV) imaging, laser reflections, and low-light level imaging. And yet again, Shackleton Crater continues to confound us with contradictory evidence, both for and against the presence of water-ice in its interior.

In 2009, the question regarding the presence of water-ice somewhere near the lunar south pole was answered when the LCROSS

Fɪɢ. **4.12** Shackleton Crater, named after Antarctic explorer Ernest Shackleton, is located at the lunar south pole. Because the Moon's axis is only slightly tilted, the crater's floor is in perpetual shade. The average temperature there is 300° below zero F. (Credit: NASA)

impactor threw up a cloud of water vapor and ice particles during its collision with the floor of the nearby crater Cabaeus. Spectral mapping instruments on three different spacecraft (Chandrayaan-1, Cassini, and EPOXI) documented the presence of adsorbed water on the lunar surface, increasing in concentration with latitude toward both poles.

A small impact probe flown by India (MIP) passed through a water vapor zone in the exosphere just above the lunar south pole. And radar images from Mini-RF, a radar imaging experiment on both Chandrayaan-1 and the LRO, found evidence of reflections (just as Clementine had suggested in 1994) within the interior of Shackleton Crater.

These new lines of supporting evidence were countered by Japanese researchers, whose Kaguya spacecraft imaged the interior of the crater and found morphology similar to other lunar craters in the same size-class. But no one had ever claimed that the interior of Shackleton was a skating rink of pure ice; the lunar polar ice is partly covered by waterless dust and mixed with an unknown amount of dry regolith.

Interpretation of the new data continues to vex scientists. The LOLA (laser altimeter) team on LRO recently published a paper that documents the high reflectivity (at 1 μm wavelength) of the walls of Shackleton. Although the team's favored interpretation is that this is caused by a constant exposure of fresh material on a steep slope, they also note that it is consistent with the presence of water-ice on the walls of the crater. In addition, a team analyzing neutron spectrometer data from both LP and LRO found evidence in the fast neutron data (never before analyzed) that water in the interior of Shackleton is a possible explanation for its signal.

Detailed analysis of the Mini-RF data for Shackleton corrected for its steep wall slopes and found that the presence of 5–10 wt% water there provides the best model fit to the observed data. Newly obtained UV images from LRO show the existence of water frost in the interiors of the craters Haworth and Shackleton, and the neutron detector on LRO shows enhanced hydrogen within both Shoemaker and Shackleton craters. The Japanese team from Kaguya continue to insist that the no-ice interpretation is the correct one.

So scientists are left with a mystery. Some evidence is pro-ice and some is contra-ice. However, for most of the investigators, new data does not necessarily change any minds, but tends to be interpreted in a way most favorable to their previously published ideas. This should not be terribly surprising; the people who have argued for some specific interpretation presumably did so for good reasons and desire hard and clear-cut evidence to the contrary before abandoning a previously held position, one no doubt reached after much thought and soul-searching.

The way to unravel the water-ice mystery is to go to the surface of the lunar south pole (or both poles) and measure the composition of the surfaces in question. Getting a definitive answer about the nature of lunar water would be game changing. Paul Spudis, Senior Staff Scientist at the Lunar and Planetary Institute, says the bigger mystery is: "Why hasn't the United States sent a rover to the south pole of the Moon to take a closer look?"

Compton-Belkovich Region

The far side of the Moon is home to a rare set of dormant volcanoes that has changed the face of the lunar surface (Fig. 4.13).

Data and photos from NASA's Lunar Reconnaissance Orbiter (LRO) reveal the presence of now-dead silicate volcanoes called

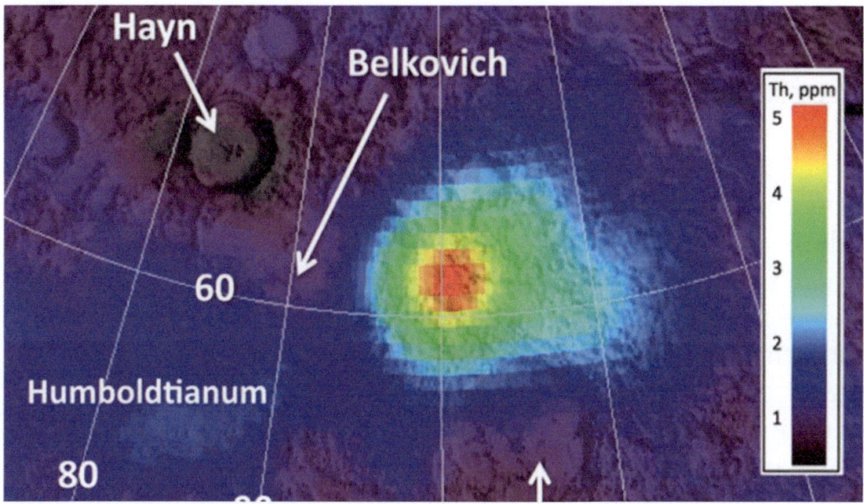

Fig. 4.13 This image from NASA's LRO shows a region on the far side of the Moon between the Compton and Belkovich craters. The colored region marks a high amount of the mineral thorium, which is thought to have been deposited by rare silicate volcanoes in the past

Compton and Belkovich, not the more common basaltic volcanoes that litter the Moon's surface, researchers said.

"Most of the volcanic activity on the Moon was basaltic," said primary author Brad Jolliff of Washington University. "Finding other volcanic types is interesting as it shows the geologic complexity and range of processes that operate on the Moon, and how the Moon's volcanism changed with time."

Lunar volcanism is very different from terrestrial volcanism because the Moon is a small body that cooled quickly and never developed rock-recycling plate tectonics like those on our planet.

The Moon, thought to have been created when a Mars-size body slammed into Earth about 4.5 billion years ago, was originally a hellish world covered by a roiling ocean of molten rock some 400 km deep. But because the Moon was small and had no atmosphere, the magma ocean cooled quickly, within perhaps 100 million years. Eventually lighter minerals such as feldspar crystallized out of the magma and floated to the top to create huge masses of feldspathic rock that formed the lunar highlands. Denser iron- and magnesium-rich minerals sank when they crystallized, forming the upper part of the Moon's mantle.

The differentiation of the crust and mantle was followed by a wave of volcanic activity between about 3–4 billion years ago, when basaltic lavas erupted on the lunar surface, filling old impact craters and other low spots to form the lunar mare.

One of the mysteries of lunar volcanism is the unequal distribution of these flood basalts. Nearly a third of the Moon's near side is covered by ancient flood basalts but the Moon's far side, where the crustal rocks are thicker, has much less.

Moreover, almost all of the volcanism on the Moon is basaltic rather than silicic, enriched in minerals containing the elements iron and magnesium rather than the elements silicon and aluminum.

Earth's continental crust, which reflects active geological processes such as subduction, magma intrusion and mountain building, includes many rocks whose compositions are intermediate between basalt and silica-rich rocks like granite, which are common on Earth. On the Moon, on the other hand, there are many basaltic rocks and only a small fraction of granite. Rocks of intermediate composition are all but missing.

Basaltic lava eruptions are generally effusive and fluid. Silicic eruptions are associated with extremely viscous lava and explosive volcanism. Non-basaltic lunar volcanism is quite rare. Examples of near side silicic volcanism have been identified in Mare Procellarum.

The Compton-Belkovich region on the northern far side (Longitude: 99.8° Latitude: 61.6°) lies between two very large ancient impact craters, Compton and Belkovich, and was found to contain large concentrations of the radioactive chemical element thorium back in 1998 when it was detected by the gamma-ray spectrometer instrument on the Lunar Prospector spacecraft. Known since as the Compton-Belkovich Thorium Anomaly (CBTA) it has remained a bit of a mystery which the Lunar Reconnaissance Orbiter is helping to unravel (Fig. 4.14).

The anomaly is between the Belkovich Crater, which is 214 km (133 miles) wide, and the Compton Crater, which is 162 km (101 miles) wide. The region as a whole is 32 km (20 miles) wide and 18 km (11 miles) long. The center of the region is a volcanic complex, 25 km (16 miles)–35 km (22 miles) across, between the Belkovich and Compton craters.

Fig. 4.14 Small dome in the Compton-Belkovich region (61.33 °N, 99.68 °E). Evidence indicates a volcanic origin for this and other intriguing features in the region

The topography of the area includes a series of domes ranging from around 1 to 6 km in diameter, some depressions of which could be evidence of collapsed volcanic features, and it is also a highly reflective region. The steeply sloping sides of some of the domes are thought to have been the product of slowly flowing viscous lava. And the fact that there are few impact craters in this area indicates that, in lunar terms, this is a fairly recent fresh feature.

Because the Moon's rotation has been affected by tidal forces between Earth and the Moon, only one side of the Moon is visible from Earth. The far side of the Moon—sometimes referred to inaccurately as the "dark side"—was hidden from view until 1959, when Soviet Union's *Luna 3* spacecraft took the first photos of the region.

However, it wasn't until the LRO captured higher-resolution images for the region that this volcanic activity could be confirmed.

The spacecraft found a number of domelike features with steeply sloping sides—telltale signs of lunar volcanoes.

Jolliff said that the domes likely formed by lava probably came from deep within the Moon. It flowed upward through cracks to pool just beneath the surface, where it pressed out to form large domes.

Lava continued to work its way to the surface throughout the area, building other, smaller volcanic domes. Some areas then collapsed, creating the irregular depressions observed by LRO's camera, researchers said. The research is detailed in the journal *Nature Geoscience*.

Most volcanoes, on Earth and off, are near other volcanoes. But the grouping in the Compton-Belkovich region is isolated. "This small volcanic complex occurs far away from the part of the Moon where most of the volcanic activity was concentrated, and where other silicic volcanism occurred," Jolliff said. "That's a puzzle."

Older, defunct volcanoes are not uncommon. Scientists have known for years that volcanoes on the Moon filled in craters to form the dark maria visible from Earth's surface. However, those lava flows are basaltic in nature.

The team also used the Diviner Lunar Radiometer Experiment to confirm the type of rocks on the plain. "Very few minerals have an infrared spectrum that can explain Diviner's observations of Compton-Belkovich and the other non-basaltic volcanoes on the Moon," explained study co-author Timothy Glotch of Stony Brook University. In fact, the rocks were silicate-rich.

"We've known for a while that the Compton-Belkovich had an unusually high thorium content," Glotch said. "Now we can positively say that that thorium is related to these silicic volcano materials."

Glotch, working with another team, was the first to identify non-basaltic volcanoes on the near side of the Moon. Due to their highly reflective surface, this group was also originally noticed by the Lunar Prospector.

However, lava from the surrounding maria may have also concealed details of the volcanoes, so some details of the region's geologic history could have been hidden, researchers said. But the volcanoes on the far side have no maria nearby to hide their features. The complete view of the volcanism in the area lies open to examination.

Fig. **4.15** Wide view of the Compton-Belkovich region

Similarly, they are surprisingly free from impact craters, which reveals a great deal about their age, researchers added (Fig. 4.15).

The early life of the Solar System was violent, with rocks scarring the surface of the planets and their moons. Features that lack this scarring formed after things had calmed down.

Jolliff and his team estimated the age of the Moon's rare far side silicate volcanoes to be about 800 million years old. Such an age would extend the volcanic activity of the Moon by 200 million years, they said.

According to Glotch, the discovery of non-basaltic volcanoes on the far side of the Moon "shows that the Moon is more compositionally diverse than we realized before this new age of lunar exploration."

"As scientists, we're still digesting all this relatively new data and working to understand what it means in terms of lunar history."

It wasn't so very long ago, Jolliff says, that scientists talked about the Moon as having two sorts of terrain, the dark maria, or "seas," and the light terra, or highlands. This simple picture of the Moon's geology served for many years, but in 2000, Jolliff and his colleagues introduced a concept in which they distinguished three different "terranes," or regions, of the Moon with distinctive geologic histories.

One of these, which encompasses much of the mare basaltic volcanism on the Moon, is called the Procellarum KREEP Terrane, or PKT for short. This immense lunar "hot spot" contains high concentrations of thorium and other radioactive, heat-producing elements, such as potassium and uranium. [KREEP stands for potassium, (K), rare-earth elements (REE), and phosphorus (P).]

As the magma ocean cooled, Jolliff explains, elements such as thorium were preferentially excluded from crystallizing minerals, forming pockets of KREEP-rich magma sandwiched between the crust and mantle.

A concentration of heat-producing elements under the Procellarum KREEP Terrane may be partly responsible for the intensive mare volcanism there. The maria, Jolliff explains, were formed when the hot radioactive elements melted minerals deep in the Moon's mantle, forming basaltic lava that erupted through fissures onto the Moon's surface. Well over half the Procellarum KREEP Terrane was resurfaced by volcanism.

Although most of the volcanism was of the basaltic variety, resulting in the large, dark patches on the Moon visible to the unaided eye from Earth, a much rarer form of volcanism, one that produced lavas rich in silica, also occurred in the PKT. These volcanic deposits are known as "red spots" because of their spectral characteristics, and results from the LRO spacecraft confirmed their silica-rich compositions. The red spots include some with distinctive dome shapes, some quite large, and all within the boundaries of the PKT.

Ever since the Lunar Prospector mission first revealed the thorium-rich bull's-eye isolated on the far side of the Moon and distant from the Procellarum KREEP Terrane, Jolliff's group has been curious as to what it was. "When the Lunar Reconnaissance Orbiter was launched in 2009, we were finally able to image it at high resolution," he says.

At the center of the thorium bull's-eye is a small volcanic complex, 25–35 km across, depending on the direction, nestled between the Compton and Belkovich craters, which are 162 km and 214 km across, respectively. Significantly, the Compton-Belkovich Thorium Anomaly lies about 900 km from the north-eastern extent of the Procellarum KREEP Terrane.

"In the initial LRO images, taken with the orbiter's Narrow-Angle Cameras (NACs) during the commissioning phase of the mission, we could see lumpy terrain and collapse features that might be volcanic," Jolliff says.

But impacts that blast material out of craters and onto the surrounding surface can also produce lumpy or mountainous deposits. "To be sure we were seeing volcanic features, we looked more closely, using images from the NACs taken during the mapping phase once the orbiter had reached its 50-km circular orbit. The NACs have telescopic optics and produce images with a 50-cm-per-pixel resolution when in a 50-km-altitude orbit.

"We mapped the same area more than once with the NACs," Jolliff says. "We went over looking straight down at the feature we were studying, and then we tilted or 'slewed' the whole spacecraft on the next orbital pass so that we could image the same feature at a different angle. From those two views we built a three-dimensional model of the terrain—a digital terrain model, or DTM—that allows us to rotate the terrain in the computer."

Among the diagnostic features revealed by these perspective views are a mountain whose features are commonly obscured by shadow, but when viewed with the terrain model, can be seen to have a depression at the summit, and what appears to be an area of the rim where a breach occurred together with subsequent collapse and mass wasting (downslope movement of rock or regolith).

The scientists bolstered the case for volcanism by taking cross sections through the terrain and measuring slopes along these cross sections. This showed, for example, that one of the domes has a summit plateau with a broad central depression, both characteristics typical of a volcano.

"It's unusual for a geologic formation to have a flat top or a depression at the top unless it's volcanic," Jolliff says. "What happens is that the magma subsides inside after the eruption, leaving a depression."

The steepness of the volcanic features provided clues to the composition of the lava that formed them. Whereas the basaltic shield volcanoes elsewhere on the Moon typically have slopes of 7° or less, the Compton-Belkovich volcanoes have slopes that reach 20–25°, suggesting they were formed by a more viscous lava.

Another piece of the puzzle came from one of LRO's other instruments, a thermal radiometer called Diviner. (This instrument records radiation in the infrared part of the spectrum, including bands sensitive to mineralogy as well as to temperature.)

Working with Ben Greenhagen, Ph.D., a former graduate student at Washington University in St. Louis and now the deputy principal investigator of the Diviner instrument, the team was able to tell there was a concentration of silica-rich rock, such as granite or rhyolite, at the Compton-Belkovich volcanic feature.

"That's very unusual," Jolliff says. "There are only about a half dozen other features on the Moon that are thought to be silica-rich, because the Moon, unlike the Earth, does not reprocess rock materials in a way that concentrates silica."

Jolliff and his team can't yet be sure and are reluctant to speculate, but they suspect the newly discovered volcanic province might be much younger than most of the volcanic features in the Procellarum KREEP Terrane.

"Although we know from direct analysis of lunar rock samples that most of mare volcanism occurred 3–4 billion years ago, we can see from orbit some mare basalt flows that might have occurred as recently as 1 billion years ago," Jolliff says.

"We don't have a way to get an absolute date on the Compton-Belkovich volcanic feature because we don't have rocks in hand," Jolliff says, "but since there are relatively few craters, the surface actually looks pretty fresh. And we see small-scale features that haven't been completely beaten up and obliterated by the impact process."

"If this volcanic province formed very late in the game, it couldn't have been due to radioactive decay because those heat sources diminish with time and it gets harder and harder to get lavas to the surface." "But," he says, "the Moon may still have a molten outer core. "But if the outer core is molten, it might generate pulses of heat. After all, on Earth we have hot spots like the Hawaiian volcanic chain that are associated with deep plumes of heat."

"A pulse of heat from deep in the mantle might melt a pocket of KREEP-rich rock at the base of the crust left by the original crystallization of the magma ocean," Jolliff says. "The melt might have risen from the base of the crust to near the surface, inflating the surface in a low dome when it ponded there."

"As this lava began to crystallize," Jolliff says, "it would have differentiated to produce a more silicic melt that was enriched in thorium. The effusive eruption of this material could have produced the elevated features to the west and east sides of the Compton-Belkovich feature and broad, low, flow features. Late eruptions of the most silicic lava would have formed domes with steep flanks. Some of the very latest eruptions might have been small bodies of very silica-rich magma that just barely breached the surface, forming small bulges."

"It might have been like the Novarupta dome in Katmai National Park in Alaska. A very viscous lava bulges up like a balloon coming up from below. It inflates, and the surface may split and crack, and then it just solidifies in place," he says, and adds "As these volcanic features pushed upward, other parts of the surface nearby may have collapsed, forming an irregularly shaped caldera in the center of the feature."

"At least that's currently my favorite hypothesis for what happened," he says smiling. "What we really need to test this and other new ideas about the Moon is sustained human exploration of our nearest and geologically very interesting neighbor in space."

Cabeus

A frigid crater at the Moon's south pole is jam-packed with water ice, with some spots wetter than Earth's Sahara desert, boosting hopes for future lunar bases.

That's the picture painted by six new studies that analyzed the intentional Moon crash of a NASA spacecraft in October 2009. The agency's LCROSS probe was looking for signs of water when it smashed into Cabeus Crater at the Moon's south pole, and the spacecraft found plenty of it.

The results expand on original findings, revealing that Cabeus harbors many other compounds, too—stuff like carbon monoxide, ammonia, methane, mercury and silver. And the new studies—reported as six separate papers in the journal *Science*—put a solid number on the amount of frozen water at the Moon's south pole. Water ice makes up about 5.6 % of the total mass on the floor of Cabeus, making the crater about twice as wet as Sahara Desert soil, according to LCROSS mission principal investigator Tony Colaprete.

"That is a surprise," said Colaprete, who works at NASA's Ames Research Center in Moffett Field, CA. "And it has a lot of ramifications in terms of our understanding of water and other volatiles on the Moon."

The high concentration of water-ice came as a bit of a shock to mission scientists.

"I still can't really wrap my brain around it," said Colaprete, who led one of the studies reported in the journal *Science* and is a co-author on several others. "There are places on the Moon that are wetter than parts of Earth—that's kind of neat." This Moon water-ice is also relatively pure, the researchers found.

The LCROSS spacecraft picked up ice signatures for four full minutes. If the ice crystals had been impregnated with lots of lunar dirt grains, that signal would have faded within 20 s or so, according to Colaprete, since grains heat up fast in sunlight.

"For ice crystals to last more than a minute, they need to be 80 or 90 % water ice," Colaprete noted. "Otherwise they will sublime, evaporate in the sunlight" (Fig. 4.16).

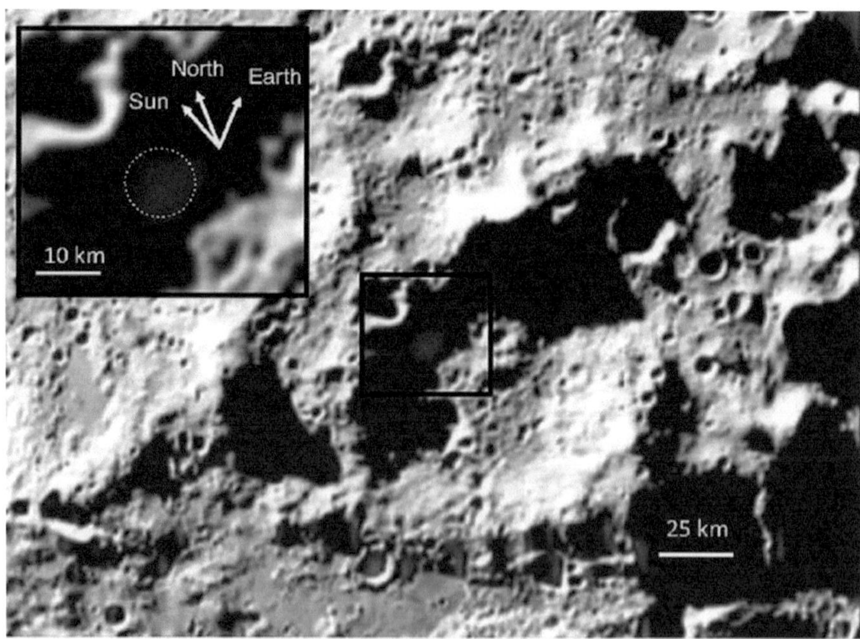

FIG. 4.16 An image of debris, ejected from the Moon's Cabeus Crater and into the sunlight, about 20 s after the Centaur rocket's impact. The *inset* shows a close-up with the direction of the Sun and Earth. (Credit: NASA)

Another intriguing result was the variety and amount of other substances inside Cabeus. LCROSS and a sister probe, the LRO, found evidence of all kinds of compounds, including elemental hydrogen, carbon monoxide, ammonia, methane, mercury, calcium, magnesium and silver. And these materials made up a surprisingly large chunk of the crater floor.

"Where we impacted, up to 20 % was something other than dirt," Colaprete said. "It was ices, volatiles, light metals. That was a surprise that you had so much of this material in there."

The LCROSS spacecraft, short for Lunar Crater Observation and Sensing Satellite, was built to live fast and die young. It launched, along with the LRO, in June 2009 aboard a Centaur rocket.

On October 9 of that year, the Centaur hurtled solo toward Cabeus, a 60-mile-wide (97-km-wide) crater near the Moon's south pole. When the rocket hit, it raised a huge debris plume up into the sunlight, where the two probes could scan it with their instruments, which included cameras and various spectrometers.

LCROSS plummeted just four minutes behind the Centaur, getting an up-close look at the ejecta cloud before smashing into the lunar surface itself. The LRO spacecraft watched all this action from above, peering at the two impacts' debris plumes. It remains in lunar orbit today, mapping the Moon's surface. Scientists announced that those plumes contained "significant amounts" of water.

Now, after analyzing more of the data gathered by both LCROSS and LRO, they have a much better idea of just what's in the bottom of Cabeus Crater—and they're gaining a better understanding of how it may have gotten there.

The original source of much of the material is likely asteroid or comet impacts, scientists said. Once they arrived, the compounds could have moved all over the lunar surface—liberated from the dirt by micrometeorite strikes or solar heating—until they hit a cold trap like Cabeus.

The permanently shadowed inside of Cabeus is among the coldest places in the solar system, with average temperatures around –387 °F (–233 °C). Many compounds would sink into these frigid depths and never surface again. So water, ammonia and everything else could keep accumulating in the crater for billions of years.

"This place looks like it's a treasure chest of elements, of compounds that have been released all over the Moon," said Peter Schultz of Brown University, lead author of one of the *Science* papers and co-author of another one. "And they've been put in this bucket in the permanent shadows."

However, Colaprete thinks there's more to the story at Cabeus.

The new research details, in part, how the crater was chosen for the LCROSS kamikaze mission: LRO instruments picked up a strong hydrogen signal in the crater, indicating the likely presence of lots of water-ice. But there are many freezing-cold craters at the Moon's south pole, and most of them didn't show such a strong hydrogen signature. And some of the places with lots of hydrogen aren't even in permanent shadow.

Cabeus stands out, indicating that there's likely more to accumulating large quantities of water—and other materials—than just frigid temperatures. "I think the best model right now, given the compounds we see, is that the Cabeus site is actually a comet impact site," Colaprete said.

That's not to suggest that volatiles don't migrate around the Moon and get trapped in the bottom of permanently shadowed craters. That's likely happening, too, Colaprete said. But that background process likely can't fully explain Cabeus.

"It seems to suggest that our old thinking about this kind of uniform emplacement of water over a billion years is only a part—and maybe a minor part—of the story when it comes to these pockets of high concentrations," Colaprete said.

The high concentration of water-ice at the bottom of Cabeus is good news for anyone pushing for bases at the Moon's poles. Future Moon-dwellers could conceivably mine such large quantities of ice efficiently. They could process it into its constituent hydrogen and oxygen, prime ingredients of rocket fuel. And they could melt the ice down and drink it—provided they remove some of the nasty stuff, like mercury.

Some of the other compounds found in the crater—such as elemental hydrogen, methane and ammonia—could be useful, too, according to Colaprete. "These places are definitely resource-rich and suggest that they would be advantageous to use for producing resources, if it ever came to that," Colaprete said.

There's no reason to think Cabeus is an anomaly, Colaprete said. There could be other super-enriched sites like it at both the north and south poles. And the poles more generally could harbor a lot of water ice, according to the new research. Modeling results support the possibility that there may be large regions of lunar "permafrost," where relatively accessible ice could be trapped below the surface, even in warmer spots that see the sun occasionally.

And that's a good thing, because Cabeus itself may not be the ideal site for a lunar base. For one thing, the floor of the crater is in permanent shadow and incredibly cold. It's tough to design equipment that can operate at temperatures of –387 °F (–233 °C)—and that equipment likely can't be solar-powered.

Also, the findings suggest that all those volatile compounds form a soft, frosty layer on the Cabeus floor that could bog down rovers or landers. When the Centaur hit, LCROSS measured a 0.3 s delay until a big heat flash resulted.

That's a very long lag, especially considering that the rocket was moving at 5580 mph (9000 kmph). The result suggests that the soil is very porous, perhaps almost fluffy.

"If we had hit rock, that flash would have happened almost instantaneously," Colaprete said.

The $79 million LCROSS mission wasn't the first to find water on the Moon—three other spacecraft had previously detected evidence of water-ice on the lunar surface, a finding announced just a few weeks before LCROSS' kamikaze plunge.

However, the LCROSS mission is still yielding new insights that should change how researchers think about the Moon, according to Colaprete. That means the spacecraft's sacrifice was worth it.

"We went out with a bang," Colaprete said, "and the return has just been phenomenal."

South Pole-Aitken Basin

The South Pole-Aitken (SPA) basin is the largest and oldest recognized impact basin on the Moon. It's diameter is roughly 2500 km or 1550 miles. The Moon's circumference is just under

FIG. 4.17 Crater Ohm resides in the wide transitional zone between the below global mean elevations characterizing the Moon's near side and the disproportionate heights of the far side, north-northeast of the south pole's Aitken basin. In this simulated low orbit view from the NASA ILIADS application Ohm is seen from the south (LROC WAC 100 m Global Mosaic on LOLA laser altimetry digital elevation model, v.2). The decline in elevation along the horizon, from west to east, is authentic. (Credit: NASA)

11,000 km, meaning the basin stretches across nearly a quarter of the Moon (Fig. 4.17).

Stratigraphic relationships show that SPA is the oldest impact basin on the Moon, but scientists are intensely interested in just how old it is. Lunar samples suggest that most of the major basins on the Moon formed around 3.9 billion years ago in a period called the late heavy bombardment. By this time most of the large debris within the Solar System should have already accreted to form the planets, so such a large number of big impacts occurring at nearly the same time may have been due to unusual gravitational dynamics in the early Solar System.

Was the impact that caused the SPA basin also a part of some cataclysmic event that occurred 3.9 billion years ago? If so, that impact is strong evidence for an extreme event that would have affected all of the terrestrial planets, including Earth at a time when life was just beginning. If the basin is much older, that may suggest that instead of a spike in the impact rate at 3.9 billion years, the number of impacts simply trailed off from a peak earlier on.

How can we find out just how old the SPA impact basin is? The best way would be to sample materials from the interior of the basin and use radiometric age-dating techniques to determine when they were last molten, as heat from the impact would have melted a large volume of material, resetting radiometric clocks. But the basin is so old that its surface has been cratered many times over, meaning that some of the rocks would have had their radiometric ages reset by these subsequent impacts. So it may be difficult to find rocks with ages that truly reflect the SPA event without careful consideration of the local geology.

Given the South Pole-Aitken's prominence at the bottom of the cratering heap, its age provides a crucial constraint on this 'late heavy bombardment' or 'lunar cataclysm.' An early date for South Pole-Aitken means a broader peak in the bombardment rate, or possibly a steady rate throughout the period. A later date speaks to cataclysm.

This is why the U. S. National Academies last year called dating South Pole-Aitken the most important goal in lunar science. Date the basin, and you test the idea of a cataclysm, with all its ensuing implications. It tells you what was happening on the Moon early on and, by inference, what was going on in the rest of the Solar System. It tells you whether the inner planets got smacked suddenly in an atmosphere-annihilating blast of impacts. And that has implications for the origins of life. Was the great bombardment so severe that it sterilized any life that had got started before then? Did it create the hellish conditions that many of the earliest life-forms seem to have endured? Could it even have moved life from one planet to the next, throwing travelers such as Dhofar 961 from surface to surface, complete with bacterial hitchhikers? (Fig. 4.18).

"It's a major unsolved problem," says Jeffrey Taylor, a geologist at the University of Hawaii. "And the Moon is the only place we can address it."

FIG. 4.18 South Pole-Aitken (SPA) Basin, the biggest and most ancient lunar basin. *Arrow* identifies the location of interesting wrinkled ridges within Aitken Crater

Look at the Moon for even a moment, and it's clear that the place has been brutalized. Yet for many years, scientists thought that its cratered surface resulted from inner turmoil rather than outer. Impact craters were mistakenly identified as volcanic calderas, the remnants of explosive eruptions. "You have no idea how pervasive this idea was—that volcanics were responsible for everything on the Moon," says Don Wilhelms, a retired geologist who was the first to map the South Pole-Aitken Basin in the early 1970s, when working at the U. S. Geological Survey in Menlo Park, CA. With Steve Squyres, Wilhelms was also the first to propose the existence of the Borealis Basin on Mars.

As the space race took off, rich new Moon maps were produced, and pioneering astrogeologists buried the volcanic theories. The big basins, they deduced, were all caused by impacts. Then, by the early 1970s, the Apollo astronauts had brought back another surprise: Moon rocks that showed that the impacts were all roughly the same age. The huge Imbrium Basin came in at an age of 3.85 billion years; nearby Nectaris, separated in the relative chronology by hundreds of substantial craters, was just 50 million years younger. Nothing was older than 4 billion years.

In 1973, Fouad Tera and his colleagues at the California Institute of Technology in Pasadena first used the term 'cataclysm' to explain the extreme pace of the impacts. "It must in any event have been quite a show from the Earth, assuming you had a really good bunker to watch from," they wrote in an abstract to that year's Lunar and Planetary Science Conference.

It wasn't just a 'show from the Earth,' though; it was the greatest show Earth itself has ever experienced—the sort of show you're lucky to come through intact. The Moon and Earth are so close to each other that whatever happened on the Moon also happened on Earth—and then some. The record has been lost on Earth because most impact craters are erased through weathering, erosion and the continuous churn of plate tectonics. That makes the Moon "a witness plate for what happened on the Earth," says David Kring, a geologist at the Lunar and Planetary Institute in Houston, Texas.

And what a bad time it was. To model what Earth went through, Kring scaled up what happened to the Moon by a factor of 13 to account for the fact that Earth is a much larger target. Given Wilhelms' estimate of 15 major lunar impact basins in the 50 million years between Nectaris and Imbrium, this meant to Kring's team that a projectile big enough to form a 20-km crater hit Earth every few thousand years.

Every million years, something would come along big enough to make a 1000-km basin. Such impacts would have vaporized Earth's oceans and steam-sterilized the surface; Kring says an atmosphere of rock vapor could linger for thousands of years after the impact.

Here's the crazy part: Kring's estimate is, in fact, very conservative. Earth's strong gravity could attract impactors at a frequency as much as 500 times higher than the Moon would. Moreover, Kring does not include in his calculations the 30 other huge basins that, according to Wilhelms, were formed after South Pole–Aitken and before Nectaris. If South Pole-Aitken turns out not to be significantly older than Nectaris, then the frequency of doomsday rocks hitting Earth rises yet higher.

"It means that the Earth was probably not a very good place to be 4 billion years ago."

What's more, Wilhelms' basin count—a baseline for many studies—is now old, and is conservative itself. Herbert Frey, at the Goddard Space Flight Center in Greenbelt, MD, recently finished a new hunt for basins, based on topographical data collected by the Clementine Moon orbiter. At the 2008 Lunar and Planetary Science Conference, Frey's team reported 92 basins bigger than 300 km across—twice as many as Wilhelms. "We've grossly underestimated the actual flux of objects that hit the Moon," Frey says. "It means that the Earth was probably not a very good place to be 4 billion years ago."

Yet some astrobiologists say that a cataclysm may have catalyzed the origin of life rather than snuffed it out. A stream of comets or asteroids hitting the planet would have brought foreign organic material to Earth. The bombardment might have pierced the crust, stirring up deep convective currents in the mantle in such a way as to establish early continental crust. And, although surface oceans might have been stripped away, subsurface water and heat could have nourished heat-loving organisms. It's probably no coincidence that in phylogenetic trees of life, the roots of the three major branches—bacteria, archaea and eukaryotes—tend to be heat-loving.

One thing the lunar rocks make clear is that the bombardment dropped off pretty quickly about 3.85 billion years ago. But what was doing the pummeling in the first place? Some have claimed it was simply the expected collisions of things left over from the formation of the Solar System, but there's no obvious way there would have been enough projectiles at the beginning to last as long as would be needed for that. Others say the cataclysm was a fresh spike of new bombardments, but what would have caused such an influx of new impactors? "Some people didn't like it because they couldn't think of a mechanism," says Wilhelms.

A few years ago, one possible explanation surfaced from researchers based in Brazil, the United States and in Nice, France. The team developed a dynamical model for the Solar System that explained why Uranus and Neptune circle the Sun farther out and more eccentrically than expected. In their model (sometimes called the Nice model, since the researchers in France were based in Nice), they started the infant Solar System with Neptune's orbit inside that of Uranus and let the clock run. Some 700 million years later, Jupiter and Saturn fell into an orbital pattern, and the resulting gravitational pull caused Uranus and Neptune to be kicked farther away from the Sun. That in turn disrupted a massive disk of icy comets in the Kuiper belt beyond Pluto, and sent them hurtling into the inner Solar System.

There's one problem with all this. The chemical composition of material within lunar craters, as well as their size distribution, matches nicely with asteroids, not comets—suggesting that asteroids were the main, or most recent, impactors during the bombardment. The Nice modelers have an answer for that: the changes in Jupiter's and Saturn's orbits may have also disrupted the Asteroid Belt between Mars and Jupiter. That could have been enough to send asteroids smashing into Earth, the Moon, and more. Korotev says he is now a believer in the lunar cataclysm, thanks in part to the Nice model.

Other work is resolving his other long-standing problem with the cataclysm hypothesis—his belief that the dating of most Moon rocks to around 3.9 billion years ago is the result of an artificial selection bias. He and some other researchers have argued that the astronauts might have simply kept picking up rocks from the Imbrium impact over and over again, and that scientists interpreted them as being from different impacts.

Marc Norman, a cosmochemist at the Australian National University in Canberra, continues to wring new dates from the Apollo collection that may counter this challenge. Norman has been looking at impact melts, the recrystallized remains that contain an isotopic record of a rock being melted by an impact. In one Apollo rock, Norman dated 21 impact melts to within a 200-million-year window. And he found that the melts fell into a number of age clusters, which he interprets as representing four different impact events. If his interpretation is correct, that would be more evidence that multiple big basins were formed within the narrow time period of the bombardment.

Other evidence is coming from meteorites, which unlike the geographically constrained Apollo rock collection are thought to have been hacked from all over the Moon. Working with Kring, Barbara Cohen, now at the Marshall Space Flight Center in Huntsville, AL, analyzed four meteorites containing impact melts representing seven to nine impact events. None of the melts, she found, was older than 3.92 billion years. Even the notorious Martian meteorite Allan Hills 84001—famous a decade ago for claims that it contained evidence of life—offers support for the lunar cataclysm theory. Parts of the 4.5-billion-year-old rock were altered in some sort of major event 3.9 billion years ago.

Still, some lunar scientists are not satisfied with Moon rocks fetched by astronauts or fallen from the sky. The best way to date South Pole-Aitken, they say, is to go there, get a rock, and date it.

Sending a simple robotic lander would be relatively cheap, but the robot's dating capabilities would not be good enough. Radioisotope dating requires a mass spectrometer, and one small enough to fly on a lander would have uncertainties of 10 %.

On a 4-billion-year-old rock, that's 400 million years—exactly the sort of error that a mission travelling to date South Pole-Aitken is supposed to dispose of, not create. "You want to be able to say that your 4.2-billion-year-old age is different from a 4.0," explains Taylor. "It requires more accuracy than we have at present."

And so a group of lunar scientists is pushing for a South Pole–Aitken sample return mission, which would be the first lunar sample return since the last Soviet Luna spacecraft returned 170 g of soil in 1976. The group, led by Brad Jolliff of Washington University in St. Louis, plan to propose a mission for the next NASA New Frontiers competition, a mission class capped at $650 million.

By 2020, if NASA's plans to return people to the Moon are realized, astronauts could already be encamped at the nearby Shackleton Crater, placing them near the rim of the South Pole-Aitken basin. The moonrise lander would collect rock and soil, and return to Earth with about a kilogram of material, Jolliff says.

Samples from a revamped moonrise mission would allow lunar scientists to date many impact-melt crystals. The oldest, and most frequent, dates should correspond to the South Pole–Aitken impact. But other impact-melt dates would undoubtedly pollute the picture, as there are half a dozen other large basins within South Pole-Aitken. And debate would continue. "Landing in the middle of a field and getting a scoop of dirt is not going to give you the answer you need," says Norman.

In fact, Norman advocates both robotic and manned missions to the Moon, saying both are needed for a balanced exploration program. "My feeling about sample return may be a little more nuanced than simply humans versus robots," he says. "Both can do the job provided we do the geologic homework, and neither will do an adequate job if we don't." Others, however, argue that the mystery of the great backside basin won't be solved until a human goes to the source and plucks a rock from within its blast shadow.

However, returning people to the Moon will cost at least $230 billion over two decades (according to the U. S. Government Accountability Office), compared with the New Frontiers $650-million cut-off. And Korotev thinks he can solve some of the mystery simply by dating the half-gram piece of the Moon he bought for $542 (plus $12 shipping and insurance). He will share his precious sliver of South Pole-Aitken with Cohen. In Huntsville, she plans to sink a diamond-tipped drill bit into Dhofar 961 and extract several cores, each as fine as a human hair.

After vaporizing the samples with a laser, she will measure argon gas that has been trapped inside the rock crystal lattice for billions of years. Counting those atoms might allow her to count back in time to the blistering crucible of the bombardment. If she's successful, she will extract a date desired by lunar scientists for many moons—a big message from a little bottle.

Giordano Bruno

High-resolution images acquired in 2008 by the Terrain Camera on board the Japanese lunar orbiting spacecraft SELENE (Kaguya) show numerous small craters on the ejecta blanket of the far side crater Giordano Bruno. The 10 m/pixel spatial resolution of the images, more than ten times higher resolution than previous image data of this area, allows unprecedented study of surface details.

A team of twelve scientists from the Institute of Space and Astronautical Science at the Japan Aerospace Exploration Agency (JAXA) and other research centers in Japan used the Terrain Camera data to determine the formation age of Giordano Bruno by the

Fig. **4.19** Southern rim of Giordano Bruno Crater seen obliquely (79°) from 53 km altitude

time-honored method of counting the craters on its continuous ejecta (Fig. 4.19).

Tomokatsu Morota and coauthors estimate that Giordano Bruno is between 1 and 10 million years old, which argues against the crater's possible formation in medieval time. The SELENE (Kaguya) data has sparked additional interest in the age of this and other young craters determined by the crater-counting method.

Sitting on the Moon at 36 N, 103 E, the bright-rayed, 22-km-diameter crater, Giordano Bruno, is just out of sight from Earth but visible in images acquired from orbit. What separates this particular crater from the untold multitude of craters and basins on the Moon is an idea, proposed by Jack Hartung some 30 years ago, that people in medieval England saw the impact that formed it. Hartung's idea is based on eyewitness accounts described in the medieval chronicle of an English monk and historian named Gervase of Canterbury. In the year 1178, on the evening of June

18, five monks from Canterbury saw (an English translation from Latin):

> now there was a bright new Moon, and as usual in that phase its horns were tilted toward the east and suddenly the upper horn split in two. From the midpoint of this division a flaming torch sprang up, spewing out, over a considerable distance, fire, hot coals, and sparks. Meanwhile the body of the Moon, which was below, writhed …[and] throbbed like a wounded snake.

Though intriguing, Hartung's hypothesis was not universally accepted. Morota and coauthors summarize some of the counter arguments, which also began to surface 30 years ago. They include the 1977 argument by H. H. Nininger and Glenn Huss (American Meteorite Lab) that what the monks actually saw was a meteor blazing through Earth's atmosphere that passed in front of their view of the Moon. They also include the 2001 work by Paul Withers (then at the University of Arizona, now at Boston University) showing the complete lack of any historical record of a huge meteor storm that would have occurred in Earth's atmosphere after the formation of Giordano Bruno crater.

With the SELENE data, Morota and colleagues sought to finally determine the formation age of Giordano Bruno by counting craters on its ejecta blanket to test the medieval formation hypothesis. This was all made possible because of the beautifully detailed views captured by the Terrain Camera of the extremely fresh features of Giordano Bruno, including its pristine rim, smooth internal melt ponds, the large number of boulders inside and outside the crater, its cratered ejecta blanket, and bright rays.

Determining the age of a planetary surface by counting craters is based on the simple idea that a newly developed surface has no impact craters and older surfaces have accumulated more craters with time. (For a short explanation of the crater counting technique visit the Planetary Science Institute web page.) Morota and colleagues counted craters in a 294 km area from the crater's rim out to a distance of one crater radius. Crater size-frequency measurements are sensitive to any contamination by secondary craters, but Morota and colleagues determined that the craters (40–200 m in diameter) in this study area are primary craters formed after Giordano Bruno and not secondary craters produced by fall-back of high-velocity ejecta blocks from other recently formed large craters. This point will be revisited later in this chapter.

Fortunately, crater frequencies on lunar surfaces can be converted to very good estimates of absolute ages because we know the absolute ages of the rocks brought back by the U. S. Apollo and unpiloted Soviet Luna missions (determined in the laboratory using (radiometric dating techniques). In fact, crater-counting studies show that Earth's Moon preserves a record of almost 4 billion years of Solar System impact history. So when did Giordano Bruno crater form? A graph is used to show isochrons, lines of equal age, for the Moon determined by the cratering chronology model developed by Gerhard Neukum (Freie Universität, Berlin, Germany) and colleagues. Based on the crater counting of Morota and colleagues, Giordano Bruno formed between 1 and 10 million years ago (Fig. 4.20).

The size-frequency distribution of these small craters led Morota and colleagues to establish an age of 4 million years for Giordano Bruno. Anything Copernican in age, or less than a billion years old, is considered young in lunar parlance. The age correlates well with other indicators of crater youth determined by other

Fig. 4.20 Crater Giordano Bruno

methods. In previous studies, researchers cited the ratio of large ray length to crater diameter as evidence of the crater's young age. More importantly, the extensive ray system of Giordano Bruno has been interpreted previously as spectrally immature. By maturity we mean how long a surface has been exposed to (and displays the effects of) space weathering; hence, an immature surface has been exposed for a shorter time than a mature surface.

Significant work has gone into using orbital multispectral data correlated with lunar sample geochemistry to determine the maturity levels of lunar surfaces. The techniques developed by Paul Lucey (University of Hawai'i) and colleagues use a comparison of Clementine 950 nm/750 nm ratio to reflectance at 750 nm to calculate the optical maturity parameter (OMAT) of surface materials. Based on work by Jennifer Grier (Planetary Science Institute, Arizona) and colleagues, young ray ejecta have higher OMAT values (~0.28 to 0.22 near the rim to lower values farther away from the rim) compared with lower OMAT values (less than ~0.15 to 0.14) for older ray ejecta. The OMAT value for Giordano Bruno ray material is ~0.27, which makes Giordano Bruno the most immature of any large far side crater. Considering the 4-million-year-old age estimate and the apparent lack of large craters in the area younger than Giordano Bruno, Morota and coauthors resolved that no crater on the Moon is related to the transient spectacle witnessed in 1178.

After carefully considering the visible freshness of the ejecta blanket, Morota and colleagues concluded that the small craters they counted on it are primary craters formed by meteorite impacts after the formation of Giordano Bruno and are not secondary craters produced by fallback of high-velocity ejecta from other recently-formed large craters. Yet, current preliminary work by researchers using Lunar Reconnaissance Orbiter Camera (LROC- narrow angle camera) images from the Lunar Reconnaissance Orbiter mission aims to determine whether secondary craters may, in fact, be affecting the crater counts. These images have 0.5 m/pixel spatial resolution compared to 10 m/pixel resolution of SELENE. Jeffrey Plescia (Johns Hopkins University, Applied Physics Lab) and colleagues report that LROC images show the Giordano Bruno ejecta blanket is not uniformly cratered, that these pockmarks do not have fresh crater geometries (including sharp raised rims) expected for young primary craters, and that many are partly buried by continuous and blocky ejecta.

Based on their observations, Plescia and colleagues suggest the craters might be secondary craters formed from the Giordano Bruno event itself. In other words, they contend that some blocks of material were ejected so high during the Giordano Bruno impact event that they fell back after the ejecta blanket was laid down. Plescia and colleagues suggest the formation age of Giordano Bruno could be substantially younger if a good portion of the craters are indeed secondaries. How much younger? If half the craters were secondaries, then the age of Giordano Bruno would be about half of the reported age, or about 2 million years. What if all the craters are secondaries that formed from the Giordano Bruno impact event? Then perhaps Giordano Bruno may be young enough that someone saw it happen.

The work on Giordano Bruno Crater and other young, fresh lunar craters will continue as high-precision data help to generate more dependable estimates of absolute age. Making comparisons of these ages derived by the crater-counting method with spectral maturity OMAT parameters will also give us a more complete understanding of recent weathering rates on the Moon. Morota and coauthors speculate it will be possible to measure directly the present-day impact flux rate by searching for and dating the small, fresh primary craters in the new lunar data sets.

Most importantly, to be completely confident in the ages of young craters, we need to collect samples from at least a few of them and determine their ages in laboratories here on Earth. The big picture, of course, looks beyond the Moon itself. Accurately defining the ages of young lunar craters, such as Giordano Bruno, is vitally important for understanding the impact flux rate over the past few million years *throughout* the inner Solar System, including on Earth.

Ina

Back in 1971, *Apollo 15* astronauts orbiting the Moon photographed something very odd. Researchers called it Ina, and it looked like the aftermath of a volcanic eruption.

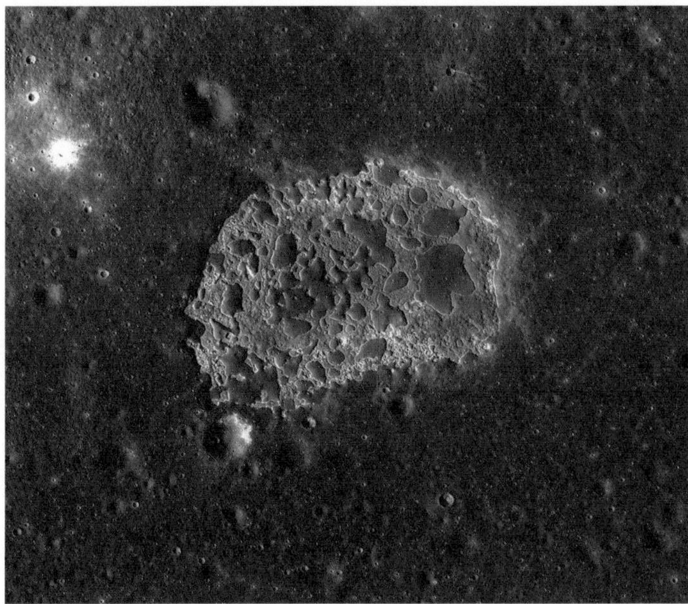

Fɪɢ. **4.21** The unusual morphology and appearance of Ina, originally referred to as 'ᴅ-Caldera' because of its unique shape, have been of interest to lunar scientists since it was first identified in Apollo images. (NASA)

There's nothing odd about volcanoes on the Moon, per se. Much of the Moon's ancient surface is covered with hardened lava. The main features of the Man in the Moon, in fact, are old basaltic flows deposited billions of years ago when the Moon was wracked by violent eruptions. The strange thing about Ina was its age (Fig. 4.21).

Planetary scientists have long thought that lunar volcanism came to an end about a billion years ago, and little has changed since. Yet Ina looked remarkably fresh. For more than 30 years Ina remained a mystery.

Turns out, the mystery is bigger than anyone imagined. Using NASA's LRO, a team of researchers has found 70 landscapes similar to Ina. They call them irregular mare patches, IMPs for short.

"Discovering new features on the lunar surface was thrilling!" says Sarah Braden of Arizona State University. "We looked at hundreds of high-resolution images, and when I found a new IMP it was always the highlight of my day."

"The irregular mare patches look so different than more common lunar features like impact craters, impact melt, and highlands material," she says. "They really jump out at you."

Located just south of the arc of the Apennine Mountains in the Moon's northern hemisphere almost midway between the craters Conon and Manilius, "Ina was first noticed by Apollo astronauts," says Schultz. It's shaped like a letter D about two km wide. Three things about Ina point to recent activity:

- Ina has mysteriously sharp edges. "Something that razor sharp shouldn't stay around long. It ought to be destroyed within 50 million years," says Braden. The destroyer of sharp edges on the Moon is a constant rain of small meteoroids that wear down mountains and craters to a nub, given time. Ina's sharp features suggest great youth.
- Ina is sparsely cratered. While small meteoroids sandblast the terrain into smoothness, larger meteoroids and asteroids make craters. The older the surface, the more heavily cratered it becomes. "Ina is almost devoid of craters," notes Braden. "We found only two clear impact craters larger than 30 m on the 8 km^2 of the structure's floor." Again, Ina appears young.
- Ina is bright and has odd colors. Rocks and dirt on the surface of the Moon grow darker as time passes. The darkening agent is space weather: a nonstop rain of cosmic rays, solar radiation and meteoroids hit the Moon and darken the ground. Ina, however, is bright, as if fresh dirt has been overturned and newly exposed. Furthermore, the colors of Ina, measured by a spectrometer on the Clementine spacecraft, are similar to the colors of the Moon's youngest craters. Yet Ina is not an impact crater.

Some of the IMPs they found are very lightly cratered, suggesting that they are no more than 100 million years old. A hundred million years may sound like a long time, but in geological terms it's just a blink of an eye. The volcanic craters LRO found may have been erupting during the Cretaceous period on Earth—the heyday of dinosaurs. Some of the volcanic features may be even younger, 50 million years old, a time when mammals were replacing dinosaurs as the dominant life-forms.

"This finding is the kind of science that is literally going to make geologists rewrite the textbooks about the Moon," says John Keller, LRO project scientist at the Goddard Space Flight Center.

Fig. 4.22 To observe Ina, find the Apennines and Conon crater on the range's southern flank, then navigate toward Manilius using high magnification. Ina appears in the *red circle*. (NASA)

IMPs are too small to be seen from Earth, averaging less than a third of a mile across in their largest dimension. That's why, other than Ina, they haven't been found before. Nevertheless, they appear to be widespread around the near side of the Moon (Fig. 4.22).

"Not only are the IMPs striking landscapes, but also they tell us something very important about the thermal evolution of the Moon," says Mark Robinson of Arizona State University, the principal investigator for the LRO's high resolution camera. "The interior of the Moon is perhaps hotter than previously thought."

"We know so little of the Moon!" he continues. "The Moon is a large mysterious world in its own right, and its only 3 days away! I would love to land on an IMP and take the Moon's temperature first-hand using a heat probe."

Some people think the Moon looks dead, "but I never thought so," says Robinson, who won't rule out the possibility of future eruptions. "To me, it has always been an inviting place of magnificent beauty, a giant magnet in our sky drawing me towards it."

Young volcanoes have only turned up the heat on the Moon's allure. Says Robinson: "Let's go!"

5. Water, Water Everywhere

Moon River

Most lunar excursions found the Moon to be a very dry world. One lunar expert reported that it was "a million times drier than the Gobi Desert." The early Apollo missions did not find even the slightest trace of water.

But after *Apollo 14*, NASA experts were stunned when a cloud of water vapor more than 100 mile^2 in size suddenly appeared on the Moon's surface. Red-faced scientists speculated that two tanks on Apollo descent stages containing between 60 and 100 lb of water became stressed and ruptured, releasing their contents.

Many scientists refused to accept this explanation, pointing out that the two tanks—from *Apollo 12* and *14*—were some 180 km apart, yet the water vapor was detected with the same flux at both sites, although the instruments faced in opposite directions. Skeptics also have understandably questioned the odds of two separate tanks breaking simultaneously and how such a small quantity of water could produce a cloud of vapor 100 miles in size that lasted 14 h. NASA later acknowledged reluctantly that the water vapor appeared to have come from the Moon's interior.

Confirmation of this came over four decades later, when NASA's Moon Mineralogy Mapper (M3) aboard India's *Chandrayaan 1* probe detected water from an unknown source deep in the lunar interior. It was the first time "magmatic water" (water derived from or existing in molten igneous rock or magma) was detected from lunar orbit, and this confirmed analyses performed on Moon rocks brought to Earth by Apollo astronauts four decades ago.

"Now that we have detected water that is likely from the interior of the Moon, we can start to compare this water with other characteristics of the lunar surface," said Rachel Klima of the Johns Hopkins University Applied Physics Laboratory in Laurel, MD, who directed a study of the M3 findings.

© Springer International Publishing Switzerland 2016

V.S. Foster, *Modern Mysteries of the Moon*, Astronomers' Universe,

DOI 10.1007/978-3-319-22120-5_5

"This internal magmatic water also provides clues about the Moon's volcanic processes and internal composition, which helps address questions about how the Moon formed, and how magmatic processes changed as it (the Moon) cooled," Klima added.

M3 imaged a 37-mile (60-km) -wide impact crater near the lunar equator called Bullialdus, whose central peak is composed of a type of rock that forms when magma is trapped deep underground. This rock was excavated and exposed by the impact that formed Bullialdus, Klima said.

"Compared to its surroundings, we found that the central portion of this crater contains a significant amount of hydroxyl—a molecule consisting of one oxygen atom and one hydrogen atom—which is evidence that the rocks in this crater contain water that originated beneath the lunar surface," she said.

The solar wind—the stream of charged particles flowing from the Sun—can create thin layers of water molecules when it strikes the lunar surface. Indeed, M3 found such water near the poles when it mapped the Moon's surface in 2009 (Fig. 5.1).

However, scientists think the solar wind can only form significant quantities of surface water at high latitudes, ruling

Fig. 5.1 Either water was inherited by the Moon from Earth during the Moon-forming impact, or it was added to the Moon later by comets or asteroids. It might also be a combination of these two processes

out this process as the source of the stuff in the more equatorial Bullialdus Crater.

Since a human first touched the Moon and brought pieces of it back to Earth, scientists have thought that the lunar surface was bone dry. But new observations from M3 and other spacecraft have put this notion to rest with what has been called "unambiguous evidence" of water on the Moon.

Liquid water cannot exist on the Moon's surface, and water vapor is decomposed by sunlight, with hydrogen quickly lost to outer space. Since the 1960s, however, scientists have speculated that water-ice could survive in cold, permanently shadowed craters at the Moon's poles.

Water (H_2O), and the chemically related hydroxyl group (–OH), can also exist in forms chemically bound to lunar minerals (rather than as free water), and evidence strongly suggests that this is indeed the case in low concentrations over much of the Moon's surface. In fact, adsorbed water is calculated to exist at trace concentrations of 10–1000 parts per million. In 1978 it was reported that samples returned by the Soviet *Luna 24* probe contained 0.1 % water by mass sample.

Inconclusive evidence of free water-ice at the lunar poles has been uncovered from a variety of observations, suggesting the presence of bound hydrogen. India's *Chandrayaan-1* detected water on the Moon and hydroxyl absorption lines in reflected sunlight in September 2009. Two months later, NASA reported that its LCROSS space probe had detected a significant amount of material in the hydroxyl group in what was thrown up from a south polar crater by an impactor. This may be attributed to water-bearing materials—what appears to be "near pure crystalline water-ice." In March 2010, it was reported that the Mini-RF on board *Chandrayaan-1* had located more than 40 permanently darkened craters near the Moon's north pole, which are believed to contain an estimated 600 million metric tons (1.3 trillion pounds) of water-ice.

Water also may have been delivered to the Moon over long timescales by the regular bombardment of water-bearing comets, asteroids, and meteoroids, or continuously produced *in situ* by the hydrogen ions (protons) of the solar wind impacting oxygen-bearing minerals.

Water inside the Moon's mantle came from primitive meteorites, new research finds, the same source thought to have supplied most of the water on Earth. The findings raise new questions about the process that formed the Moon.

The Moon is thought to have formed from a disk of debris left when a giant object hit Earth 4.5 billion years ago, very early in Earth's history. Scientists have long assumed that the heat from an impact of that size would cause hydrogen and other volatile elements to boil off into space, meaning that the Moon must have started off completely dry. But recently, NASA spacecraft and new research on samples from the Apollo missions have shown that the Moon actually has water, both on its surface and beneath.

By showing that water on the Moon and on Earth came from the same source, this new study offers yet more evidence that the Moon's water has been there all along. "The simplest explanation for what we found is that there was water on the proto-Earth at the time of the giant impact," said Alberto Saal from Brown University. "Some of that water survived the impact, and that's what we see in the Moon."

To find the origin of the Moon's water, Saal and his colleagues looked at melt inclusions found in samples brought back from the Apollo missions. Melt inclusions are tiny dots of volcanic glass trapped within crystals called olivine. The crystals prevent water escaping during an eruption and enable researchers to get an idea of what the inside of the Moon is like (Fig. 5.2).

Research from 2011 led by Erik Hauri of the Carnegie Institution of Washington, D. C., found that the melt inclusions have plenty of water—as much water, in fact, as lavas forming on Earth's ocean floor. This study aimed to find the origin of that water. To do that, Saal and his colleagues looked at the isotopic composition of the hydrogen trapped in the inclusions. "In order to understand the origin of the hydrogen, we needed a fingerprint," Saal said. "What is used as a fingerprint is the isotopic composition."

Using a Cameca NanoSIMS 50L multicollector ion microprobe at Carnegie, the researchers measured the amount of deuterium in the samples compared to the amount of regular hydrogen. Deuterium is an isotope of hydrogen with an extra neutron. Water molecules originating from different places in the Solar System have

Fɪɢ. **5.2** Data from NASA's Moon Mineralogy Mapper (M3) aboard the Indian *Chandrayaan-1* probe has shown that there is water locked in mineral grains on the Moon's surface. Scientists had previously thought that the small amounts of moisture they could detect were all being generated by the solar wind and other external factors, but the latest findings are strong evidence that the Moon contains large quantities of its own "magmatic water" from deep within its core

different amounts of deuterium. In general, things formed closer to the Sun have less deuterium than things formed farther out.

Saal and his colleagues found that the deuterium/hydrogen ratio in the melt inclusions was relatively low and matched the ratio found in carbonaceous chondrites, meteorites originating in the Asteroid Belt near Jupiter and thought to be among the oldest objects in the Solar System. That means that the source of the water on the Moon is primitive meteorites, not comets, as some scientists had thought.

Comets, like meteorites, are known to carry water and other volatiles, but most comets are formed in the far reaches of the Solar System in a region called the Oort Cloud. Because they formed so far from the Sun, they tend to have high deuterium/hydrogen ratios—much higher ratios than in the Moon's interior, where the samples in this study came from.

"The measurements themselves were very difficult," Hauri said, "but the new data provide the best evidence yet that the carbon-bearing chondrites were a common source for the volatiles in the Earth and Moon, and perhaps the entire inner Solar System."

Recent research, Saal said, has found that as much as 98 % of the water on Earth also comes from primitive meteorites, suggesting a common source for water on Earth and water on the Moon. The easiest way to explain that, Saal said, is that the water was already present on the early Earth and was transferred to the Moon.

The finding is not necessarily inconsistent with the idea that the Moon was formed by a giant impact with the early Earth, but it does present a problem. If the Moon is made from material that came from Earth, it makes sense that the water in both would share a common source. However, there's still the question of how that water was able to survive such a violent collision.

"The impact somehow didn't cause all the water to be lost," Saal said. "But we don't know what that process would be." It suggests, the researchers say, that there are some important processes scientists don't yet understand about how planets and satellites are formed.

"Our work suggests that even highly volatile elements may not be lost completely during a giant impact," said Van Orman. "We need to go back to the drawing board and discover more about what giant impacts do, and we also need a better handle on volatile inventories in the Moon."

The search for water has attracted considerable attention and prompted several recent lunar missions, largely because of water's usefulness in rendering long-term lunar habitation feasible.

Lunar scientists had discussed the possibility of water repositories for decades. They are now increasingly confident "that the decades-long debate is over" one report says. "The Moon, in fact, has water in all sorts of places; not just locked up in minerals, but scattered throughout the broken-up surface, and, potentially, in blocks or sheets of ice at depth." The results from the Chandrayaan mission are also "offering a wide array of watery signals."

The Chandrayaan mission was just one of many programs aimed at finding water on the Moon. We'll discuss these below.

Missions and Programs to Find Water

The Apollo Program

Although trace amounts of water were found in lunar rock samples collected by Apollo astronauts, this was assumed to be a result of contamination, and the majority of the lunar surface was generally assumed to be completely dry. However, a 2008 study of lunar rock samples revealed evidence of water molecules trapped in volcanic glass beads.

The first direct evidence of water vapor on the Moon was obtained by the *Apollo 14's* ALSEP Suprathermal Ion Detector Experiment, SIDE, on March 1971. A series of bursts of water vapor ions was observed by the instrument's mass spectrometer at the lunar surface near the *Apollo 14* landing site.

Luna 24

In February 1978 Soviet scientists M. Akhmanova, B. Dement'ev, and M. Markov of the Vernadsky Institute of Geochemistry and Analytic Chemistry published a paper claiming detection of water. Their study showed that the samples returned to Earth by the 1976 Soviet probe *Luna 24* contained about 0.1 % water by mass, as seen in infrared absorption spectroscopy (at about $3\,\mu m$ wavelength), a detection level about ten times above the threshold.

The Clementine Mission

Possible evidence of water-ice on the Moon came in 1994 from the U. S. Clementine probe. In an investigation known as the "bistatic radar experiment," Clementine used its transmitter to beam radio waves into the dark regions of the south pole of the Moon. Large dish antennas of the Deep Space Network on Earth detected echoes of these waves. The magnitude and polarization of these echoes was consistent with an icy rather than rocky surface, but the results were inconclusive, and their significance has been questioned.

Earth-based radar measurements were also used to identify the areas that are in permanent shadow and hence have the potential to harbor lunar ice. Estimates of the total extent of shadowed areas poleward of 87.5° latitude are 1030 and 2550 km^2 for the north and south poles, respectively. Later computer simulations encompassing additional terrain suggested that a total area up to 14,000 km^2 might be in permanent shadow.

Lunar Prospector

In 1998, the Lunar Prospector probe used a neutron spectrometer to measure the amount of hydrogen in the lunar regolith near the polar regions. It was able to determine hydrogen abundance to within 50 ppm and detected enhanced hydrogen concentrations at the north and south poles. These were interpreted as indicating significant amounts of water-ice trapped in permanently shadowed craters, but could also be due to the presence of the hydroxyl radical ($^{\bullet}$OH) chemically bound to minerals. Based on data from Clementine and Lunar Prospector, NASA scientists have estimated that if surface water-ice is present, the total quantity could be of the order of 1–3 km^3.

In July 1999, at the end of its mission, the Lunar Prospector probe was deliberately crashed into Shoemaker Crater, near the Moon's south pole, in the hope that detectable quantities of water would be liberated. However, spectroscopic observations from ground-based telescopes did not reveal the spectral signature of water.

Cassini-Huygens

More suspicions about the existence of water on the Moon were generated by inconclusive data produced by Cassini-Huygens mission, which passed by the Moon in 1999.

Deep Impact

In 2005, observations of the Moon by the Deep Impact spacecraft produced inconclusive spectroscopic data suggestive of water on the Moon. In 2006, observations with the Arecibo planetary radar showed that some of the near-polar Clementine radar returns, previously claimed to be indicative of ice, might instead be associated with rocks ejected from young craters.

If true, this would indicate that the neutron results from Lunar Prospector were primarily from hydrogen in forms other than ice, such as trapped hydrogen molecules or organics. Nevertheless, the interpretation of the Arecibo data do not exclude the possibility of water-ice in permanently shadowed craters. In June 2009, NASA's Deep Impact spacecraft, now re-designated EPOXI, made further confirmatory bound hydrogen measurements during another lunar flyby.

Kaguya

As part of its lunar mapping program, Japan's Kaguya probe, launched in September 2007 for a 19-month mission, carried out gamma ray spectrometry observations from orbit that can measure the abundances of various elements on the Moon's surface.

The Kaguya probe's high resolution imaging sensors failed to detect any signs of water-ice in permanently shaded craters around the south pole of the Moon, and it ended its mission by crashing into the lunar surface in order to study the ejecta plume content.

Chang'e 1

The People's Republic of China's *Chang'e 1* orbiter, launched in October 2007, took the first detailed photographs of some polar areas where water-ice is likely to be found.

Lunar Reconnaissance Orbiter

The search for lunar ice continued with NASA's LRO/LCROSS mission, launched on June 18, 2009. The LRO's onboard instruments carried out a variety of observations to help provide further evidence of water. On October 9, 2009, the Centaur upper stage of its Atlas V carrier rocket was directed to impact Cabeus Crater, followed shortly by the LCROSS spacecraft that flew into the ejecta plume and attempted to detect the presence of water vapor in the debris cloud.

Analysis of the data obtained from the ejecta plume confirmed the spectral signature of water. However, what was actually detected was the chemical group hydroxyl (OH), which is suspected to be from water, but could also be hydrates, which are inorganic salts containing chemically bound water molecules. The nature, concentration, and distribution of this material required further analysis.

Chief mission scientist Anthony Colaprete has stated that the ejecta appears to include a range of fine-grained particulates of near pure crystalline water-ice. A later definitive analysis found the concentration of water to be 5.6 ± 2.9 % by mass. The Mini-RF instrument on LRO observed the LCROSS landing site, but did not detect any evidence of large slabs of water-ice, so the water is most likely present as small pieces of ice mixed in with the lunar regolith. The LRO's laser altimeter's examination of the Shackleton Crater at the south pole suggests that up to 22 % of the surface of that crater is covered in ice.

Melt Inclusions in Apollo 17 Samples

In May 2011, 615–1410 ppm water was discovered in melt inclusions from the famous high-titanium "orange glass soil" sample collected during the *Apollo 17* mission in 1972. The inclusions were formed during explosive eruptions on the Moon approximately 3.7 billion years ago.

This concentration is comparable with that of magma in Earth's upper mantle. While of considerable selenological interest, this announcement gives little comfort to would-be lunar colonists. The sample originated many km below the surface, and the inclusions are so difficult to access that it took 39 years to detect them with a state-of-the-art ion microprobe instrument.

A Possible Water Cycle on the Moon

Production

Lunar water has two potential origins: water-bearing comets (and other bodies) striking the Moon, and *in situ* production. It has been theorized that the latter may occur when hydrogen ions (protons) in the solar wind chemically combine with the oxygen atoms present in the lunar minerals (oxides, silicates, etc.) to produce small amounts of water trapped in the minerals' crystal lattices or as hydroxyl groups, potential water precursors. (Mineral-bound water, or hydroxylated mineral surfaces, are not the same as water-ice.)

Trapping

Solar radiation would normally strip any free water or water-ice from the lunar surface, splitting it into its constituent elements, hydrogen and oxygen, which then escape into space. However, because of the only very slight axial tilt of the Moon's spin axis to the ecliptic plane (1.5°), some deep craters near the poles never receive any sunlight, and are permanently shadowed (Shackleton and Whipple craters). The temperature in these regions never rises above about 100 K (about −170 °C), and any water that eventually ended up in these craters could remain frozen and stable for extremely long periods of time—perhaps billions of years, depending on the stability of the orientation of the Moon's axis.

Transport

Although free water cannot exist in illuminated regions of the Moon, any such water produced there by the action of the solar wind on lunar minerals might, through a process of evaporation and condensation, migrate to permanently cold polar areas and accumulate there as ice. This is in addition to any ice brought there by comet impacts.

The hypothetical mechanism of water transport/trapping remains unknown. Indeed lunar surfaces directly exposed to the solar wind where water production occurs are too hot to allow trapping by water condensation, while no water production is expected in the cold areas not directly exposed to the Sun.

Given the expected short lifetime of water molecules in illuminated regions, a short transport distance would in principle increase the probability of trapping. In other words, water molecules produced close to a cold, dark polar crater should have the highest probability of surviving and being trapped.

Uses for Water on the Moon

The presence of large quantities of water on the Moon would be an important factor in rendering lunar habitation cost-effective, since transporting water (or hydrogen and oxygen) from Earth would be prohibitively expensive. If future investigations find the quantities to be particularly large, water-ice could be mined to provide liquid water for drinking and plant propagation. The water could also be split into hydrogen and oxygen by solar panel-equipped electric power stations or a nuclear generator, providing breathable oxygen as well as the components of rocket fuel. The hydrogen component of the water-ice could also be used to draw out the oxides in the lunar soil and harvest even more oxygen.

Analysis of lunar ice would also provide scientific information about the impact history of the Moon and the abundance of comets and asteroids in the early inner Solar System.

Ownership of the Water

The discovery of usable quantities of water on the Moon may raise legal questions about who owns the water and who has the right to exploit it. The United Nations' Moon Treaty does not prevent the exploitation of lunar resources, but does prevent the appropriation of the Moon by individual nations and is generally interpreted as barring countries from claiming ownership of *in situ* resources.

However most legal experts agree that the ultimate test of the question will arise through precedents of national or private activity. Some private companies such as Shackleton Energy Company are already asserting their right to own whatever resources they remove from the Moon or asteroids through their own effort, risk, and investment. The Moon Treaty specifically stipulates that an "international regime" is to govern the exploitation of lunar resources, but none of the major space-faring nations have ratified the treaty.

Regional Variations in Sources of Water

Scientists recently discovered that the amount of water in the Moon's rocky interior actually varies regionally—revealing clues about how water originated and was redistributed on the lunar surface.

That's not to say there's liquid on the Moon, but rather water trapped in volcanic glasses or chemically bound in mineral grains inside lunar rocks. Yet some of these rocks—those originating in the lunar interior—contain much more water than rocks found in other places. In addition, the hydrogen isotopic composition of this lunar water varies from region to region (Fig. 5.3).

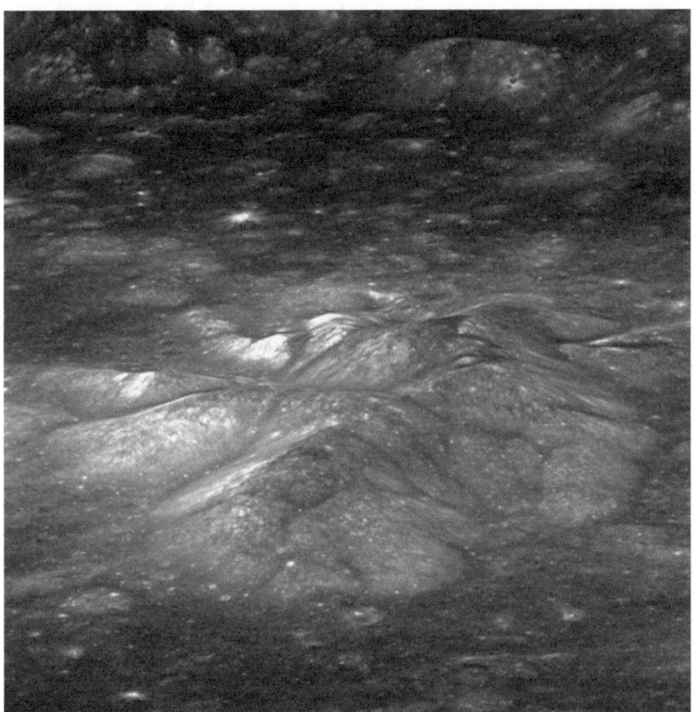

FIG. 5.3 Scientists have learned that the lunar impact crater Bullialdus has significantly more hydroxyl—a molecule consisting of one oxygen atom and one hydrogen atom—compared to its surroundings. Pictured here is the central peak of Bullialdus rising above the crater floor with the crater wall in the background

Understanding where lunar water originated from may even help scientist uncover the source of water on Earth. There are currently two theories: either the Moon inherited water from Earth during the Moon-forming impact, or it was added to the Moon later by comets or asteroids. It might also be a combination of these two processes.

"Basically, whatever happened to the Moon also happened to the Earth," Katharine Robinson, lead author of the study and graduate assistant at the University of Hawai'i—Mānoa (UHM) School of Ocean and Earth Science and Technology, said. To better understand this lunar history, researchers compiled water measurements from lunar samples. Specifically, they measured hydrogen and its isotope, deuterium, with ion microprobes. The findings brought some new clues about the water to the surface.

"This was consistent with the idea that blossomed during the Origin of the Moon conference in Kona in 1984—that the Moon formed by a giant impact with the still-growing Earth, leading to extensive loss of volatile chemicals," Robinson said. "Our work is surprising because it shows that lunar formation and accretion were more complex than previously thought."

Researchers plan to study a new set of Apollo samples from NASA, hopefully uncovering additional clues about the early life of Earth and the Moon.

Solar Photons Driving Water from the Moon

Water is thought to be embedded in the Moon's rocks or, if cold enough, "stuck" on their surfaces. It's predominantly found at the poles. But scientists probably won't find it intact while sunlight is hitting the lunar surface.

New research indicates that ultraviolet photons emitted by the Sun are likely to either quickly desorb or break apart H_2O molecules. The fragments of water may remain on the lunar surface, but the presence of useful amounts of water on the sunward side is not likely.

The Georgia Tech team, led by Thomas Orlando, a professor in the School of Chemistry and Biology, built an ultra-high vacuum system that simulates conditions in space, then performed the first-ever reported measurement of the water photodesorption cross section from an actual lunar sample. The machine zapped a small piece of the Moon with ultraviolet (157 nm) photons to create excited states and watched what happened to the water molecules.

"Overall, the Moon will lose water efficiently when the solar photons are hitting it," said Orlando. "The water also desorbs thermally. When they photodesorb or thermally desorb, the velocities are too low for the water to escape, so it will bounce around until it gets trapped in the permanently shadowed regions and the poles or break apart in transit."

According to the team's measurements, approximately one in every 1000 molecules leave the lunar surface simply due to absorption of UV light.

Georgia Tech's cross section values can now be used by scientists attempting to find water throughout the solar system and beyond.

"The cross section is an important number planetary scientists, astrochemists, and the astrophysics community need for models regarding the fate of water on comets, moons, asteroids, other airless bodies and interstellar grains," said Orlando.

The number is relatively large, which establishes that solar UV photons are likely removing water from the Moon's surface. This research, which was carried out primarily by former Georgia Tech Ph.D. student Alice DeSimone, indicates that the cross sections increase even more with decreasing water coverage. That's why it's not likely that water remains intact as H_2O on the sunny side of the Moon. Orlando compares it to sitting outside on a summer day.

"If a lot of sunlight is hitting me, the probability of me getting sunburned is pretty high," said Orlando. "It's similar on the Moon. There's a fixed solar flux of energetic photons that hit the sunlit surface, and there's a pretty good probability they remove water or damage the molecules."

The result, according to Orlando, is the release of molecules such as H_2O, H_2, and OH as well as the atomic fragments H and O.

The research is published in two companion articles in the *Journal of Geophysical Research: Planets*. The first discusses the water photodesorption. The second paper details the photodissociation of water and the $O\,(^3P_J)$ formation on lunar impact melt breccia.

Microwaves Could Extract Water from the Moon

The next time astronauts land on the Moon, it's possible they will be visiting an outpost where they can pick up some fuel and a refreshing container of liquid.

That outpost won't be offering the 64-oz Big Gulp soft drinks that you find at many of the convenience stores across the country, but it will be offering a critical commodity—water (Fig. 5.4).

Research conducted by material scientists may lead to the ability to extract water from the Moon by shooting microwave beams into the surface, according to Bill Kaukler, an Associate Research Professor in the Center for Materials Research at the University of Alabama in Huntsville.

"A lot of people think that water doesn't exist on the Moon," said Kaukler. "It's true that not all parts of the Moon have water. Where the Apollo missions landed, there isn't much water because it is exposed to the Sun half of the time. However, in the polar regions, exploratory satellites have found huge amounts of hydrogen, which is evidence that water exists."

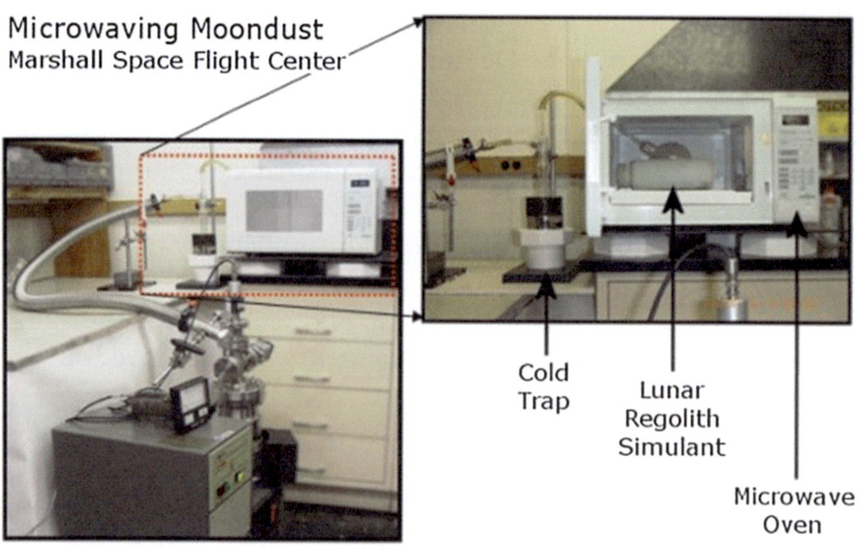

FIG. 5.4 Ed Ethridge's experimental setup at the Marshall Space Flight Center in Huntsville, AL. (Credit: NASA)

Kaukler has performed research with NASA for more than 25 years and for the past 3 years has been investigating the use of microwaves to replenish water on space missions or as a rocket fuel supply.

The Moon's surface is covered with over 2 m deep of regolith (like soil), a layer of loose, powdery, heterogeneous material created by hundreds of millions of years of meteorite and comet bombardment. Below that covering lies bedrock. "Ice is just inches below the surface of the Moon in craters at the poles [where solar heating doesn't occur]," he said.

Kaukler and Marshall Space Flight Center scientist Edwin Ethridge have been conducting research on the use of microwaves to warm the lunar regolith to draw the water up to the surface.

"Using microwaves to heat the soil offers several advantages," Kaukler said. "Microwaves are not strongly absorbed by the regolith (soil) so they can penetrate several feet into the soil and heat it." Heating is possible because the Moon's soil has about 5 % iron, similar to volcanic rock on Earth, according to Kaukler. Microwave absorption is the most efficient method to heat large volumes of regolith or rock.

He said research shows that if the regolith can be warmed from –150 to –50 °C, the vapor pressure of the water mixed in with the regolith particles is much higher than the Moon's atmospheric pressure. Kaukler said that the Moon's vacuum environment literally percolates the water vapor to the surface through the regolith particles. The water vapor collects on a cold (below –50 °C) plate where it forms as ice and is scraped off for human consumption or where it can be converted by electrolysis to hydrogen and oxygen to be used as a fuel and oxidizer that can be used in space travel, such as for going to Mars.

Kaukler, Ethridge, and other materials scientists have developed a prototype and have used simulated lunar regolith to test their ideas. Their prototype has the power of one kilowatt, about the same as a typical home microwave oven.

What their experiments show is that they are able to remove 99 % of water-ice through sublimation, or converting the frozen water directly into a gas, and could capture 95 % of the liberated water.

While the 1-kW device may prove the concept, Kaukler said a 10-kW unit would speed up the process of collecting water and make it more effective on the Moon's surface. He envisions a robotic, roving device powered by a nuclear generator to roam the Moon's surface in search of water sources.

Kaukler believes the concept of shooting microwave beams into the surface of the Moon to extract water offers several distinct advantages: not having to dig the regolith and put it into a furnace is a big advantage since heavy equipment won't be needed; leaving the Moon essentially undisturbed is important; not worrying about the underlying geology is practical since hidden or buried rocks (that have no water) could damage digging equipment.

Perhaps the most important factor of this project is not carrying water on a journey, thus saving space and weight on long-distance trips. "This is the essence of the concept of In Situ Resource Utilization or ISRU," Kaukler said. "The philosophy is to use what is on the Moon (or Mars) to make habitats without having to bring the material from Earth."

Another crucial factor is safety. The microwave process penetrates into the surface at least 2 m deep, thus eliminating the need to dig into the surface to get to the ice, according to the scientists. Kaukler said the idea of kicking up dust by digging on the Moon poses a problem for equipment and astronauts if the abrasive dust finds its way into the wrong places, as it did for the Apollo astronauts.

Research continues, according to Kaukler. More investigation is necessary to learn about the electromagnetic properties of regolith in the various microwave frequencies. Altering those frequencies could allow the device to penetrate deeper into the surface if necessary to reach additional water.

Kaukler and Ethridge are also exploring the use of microwaves for melting or sintering the regolith to make structures on the Moon.

An important concept they developed was the use of microwave melting of the regolith surface to make a dust-free crust. Such surfaces have uses as a landing pad (no dust kicked around the landing site by rockets), working surfaces to stabilize equipment and floors or roadways for astronauts to live and travel on. Conventional bricks, blocks, or walls can also be prepared this way without bringing adhesives or special cements to the Moon.

How the Moon Produces Its Own Water

The Moon is a big sponge that absorbs electrically charged particles given out by the Sun. These particles interact with the oxygen present in some dust grains on the lunar surface, producing water. This discovery, made by the ESAISRO instrument SARA onboard the Indian *Chandrayaan-1* lunar orbiter, confirms how water is likely being created on the lunar surface.

It also gives scientists an ingenious new way to take images of the Moon and any other airless body in the Solar System.

The lunar surface is a loose collection of irregular dust grains, known as regolith. Incoming particles should be trapped in the spaces between the grains and absorbed. When this happens to protons they are expected to interact with the oxygen in the lunar regolith to produce hydroxyl and water. The signature for these molecules was recently found and reported by *Chandrayaan-1's* Moon Mineralogy Mapper (M3) instrument team.

The SARA results confirm that solar hydrogen nuclei are indeed being absorbed by the lunar regolith but also highlight a mystery: not every proton is absorbed. One out of every five rebounds into space. In the process, the proton joins with an electron to become an atom of hydrogen. "We didn't expect to see this at all," says Stas Barabash, Swedish Institute of Space Physics, who is the European Principal Investigator for the Sub-keV Atom Reflecting Analyzer (SARA) instrument, which made the discovery (Fig. 5.5).

Although Barabash and his colleagues do not know what is causing the reflections, the discovery paves the way for a new type of image to be made. The hydrogen shoots off with speeds of around 200 km/s and escapes without being deflected by the Moon's weak gravity. Hydrogen is also electrically neutral, and is not diverted by the magnetic fields in space. So the atoms fly in straight lines, just like photons of light. In principle, each atom can be traced back to its origin, and an image of the surface can be made. The areas that emit the most hydrogen will show up the brightest.

Although the Moon does not generate a global magnetic field, some lunar rocks are magnetized. Barabash and his team are currently making images, to look for such 'magnetic anomalies' in

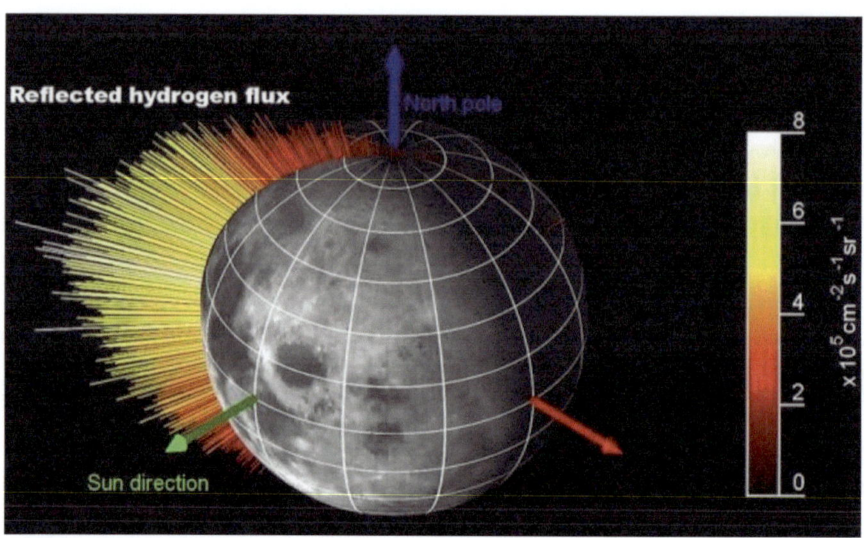

FIG. 5.5 *Chandrayaan-1* SARA measurements of hydrogen flux recorded on the Moon on February 6, 2009. (Credit: NASA)

lunar rocks. These generate magnetic bubbles that deflect incoming protons away into surrounding regions, making magnetic rocks appear dark in a hydrogen image.

The incoming protons are part of the solar wind, a constant stream of particles given off by the Sun. They collide with every celestial object in the Solar System but are usually stopped by the body's atmosphere. On bodies without such a natural shield, for example, asteroids or the planet Mercury, the solar wind reaches the ground. The SARA team expects that these objects, too, will reflect many of the incoming protons back into space as hydrogen atoms.

This knowledge provides timely advice for the scientists and engineers who are readying ESA's BepiColombo mission to Mercury. The spacecraft will be carrying two similar instruments to SARA and may find that the innermost planet is reflecting more hydrogen than the Moon because the solar wind is more concentrated closer to the Sun.

6. Shake and Quake Riddles

Blasts from the Past

Unlike Earth, the Moon has no active volcanoes despite the fact that recent moonquake data suggest that there is a lot of magma under its surface.

A team of Dutch earth scientists—led by Mirjam van Kan Parker and Wim van Westrenen from VU University in Amsterdam—have now identified the likely reason for the lack of volcanic activity: the hot, molten rock contained within the Moon's interior could be so dense that it is too heavy to rise to the surface. The team came to this surprising conclusion by producing microscopic copies of Moon rock collected by the Apollo missions and melting them at extremely high pressures and temperatures, similar to those found inside the Moon. They then measured their densities with powerful X-rays generated at the European Synchrotron Radiation Facility (ESRF).

Using synthesized Moon rock, they applied pressures of 4.5 GPa and temperatures of 1500 K in order to replicate the conditions found in the Moon's core. They did this by heating them with a powerful electric current while squashing them in a press. They then used a powerful synchrotron X-ray beam at the ESRF to measure the density of the Moon rock when solid and molten. These measurements were then combined with computer simulations to calculate the magma density at any location in the Moon.

Most lunar magmas were found to be less dense than their solid surroundings, just as they are on Earth. However, some samples from the *Apollo 14* mission, which was comprised of titanium-rich glass, were found to produce a magma as dense as the rocks found in the deepest part of the lunar mantle today. This magma wouldn't move towards the surface (Fig. 6.1).

This titanium-rich magma could have only been formed by melting titanium-rich solid rocks. Previous experiments have shown that these sorts of rocks were formed shortly after the for-

© Springer International Publishing Switzerland 2016

V.S. Foster, *Modern Mysteries of the Moon*, Astronomers' Universe,

DOI 10.1007/978-3-319-22120-5_6

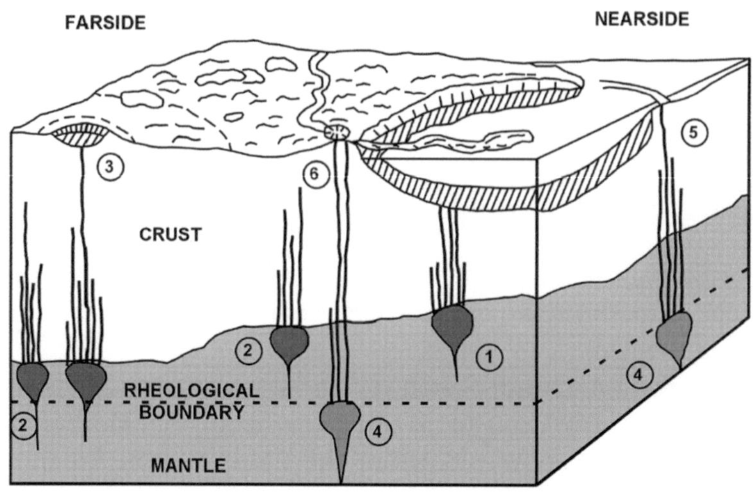

Fig. 6.1 The density of the lunar crust is very low, so hot magma is not buoyantly stable within this crust. Just like water under ice, the neutral buoyancy zone of magma is just below the lunar crust (which is about 40 km thick). So, that is where the magma stays and collects, deep in the lunar interior. (Credit: NASA)

mation of the Moon at shallow levels. It is thought that vertical movements must have occurred early in the history of the Moon, which caused these titanium-rich rocks to descend into the Moon's core-mantle boundary.

The team hasn't ruled out volcanoes emerging on the Moon's surface in the future. Wim van Westrenen explains: "Today, the Moon is still cooling down, as are the melts in its interior. In the distant future, the cooler and therefore solidifying melt will change in composition, likely making it less dense than its surroundings. This lighter magma could make its way again up to the surface, forming an active volcano on the Moon—what a sight that would be!—but for the time being, this is just a hypothesis to stimulate more experiments."

However, some scientists believe that there may be more to it than this. Our Moon's near side may be holding onto some deep-seated volcanic secrets. At least that's the tantalizing lunar take-away after 3 years of ground-based surface monitoring in visible white light.

The monitoring, done from New York City and a remote mountaintop in Chile, have produced several candidate minutes-long episodes of observed brightening that could signal that the Moon is still volcanically active. That's according to Columbia University astrophysicist and lunar scientist Arlin Crotts, the principal investigator for two optical lunar monitors, operating both on a rooftop at Columbia University and at Cerro Tololo observatory in Chile. At each location, telescopic cameras have taken hundreds of thousands of visible light lunar images at intervals of every 20 s.

Although Lunar Transient Phenomena (LTP)s have been reported by both amateur and professional astronomers for at least the past 450 years, Crotts project represents the most comprehensive scientific attempt to document the phenomena.

"We're just taking normal white light images of the whole near side hemisphere and running it through software that looks for changes between pictures," said Crotts. "There are candidate events that look very interesting. They are just small areas of the Moon that get brighter for not too many exposures. They may be outgassing events."

Crotts says that he and colleagues are only taking white light images at the moment, in order not to prejudge the phenomena's output spectrums. He adds that their software does a good job of eliminating any expected source that isn't transient to a high degree of certainty.

"You take all the hundreds of reported lunar optical transients and run them through a statistical sieve to look for consistent behavior regardless of who is observing the Moon," said Crotts. "There are just these certain areas on the Moon that always seem to be coming through."

Aristarchus plateau is the primary location for these transient events, says Crotts, followed by the craters of Plato, Grimaldi, Kepler, Copernicus, and Tycho. Historically, such putative events have been described as bright flashes, exhibiting glowing reddish and blue hues, sometimes even cast with a fog-like hazy pall.

What is known is that 3.5 billion years ago, Crotts says, the Moon regularly spewed forth large amounts of water vapor and sulfur dioxide through volcanic surface vents. In fact, he says it's thought that there were small amounts of lunar volcanism up until a billion years ago.

More recently, Japan's Kaguya lunar mission detected radioactive alpha particles from radon outgassing while flying over Aristarchus. Crotts says both NASA's Lunar Prospector spacecraft and *Apollo 15* also detected outgassing at Aristarchus; however, both *Apollo 15* and Lunar Prospector's alpha particle detectors observed two other events—respectively, at the Grimaldi and Kepler craters.

Crotts says it's possible that the gas is just being produced by radioactive decay and simply using the Moon's ancient lunar plumbing to get out. He says it's "tempting" to think these random outgassing events are tied to some sort of extant lunar volcanism. They tend to stick very closely to the boundary between the lunar maria and the lunar highlands. He notes that's also where the optical transient events tend to occur.

"Where these phenomena have been reported over the last four centuries," said Crotts, "also corresponds very closely to recently identified volcanic vents which formed about 3 billion years ago. That's rather significant." Crotts says he hopes to have an LTP paper in the works in the near future. However, he's not sanguine about getting much help from the broader lunar science community.

"Lunar scientists at the end of the Apollo era decided that the Moon is dry and dead," said Crotts. "But this is not some optical illusion. Optical illusions don't make radon gas spew from the lunar surface."

Volcanism was once a major geologic force on the Moon. Elements such as uranium, potassium, and thorium reheated areas of the lower crust and upper mantle, which created a series of melts. The melts were less dense than the surrounding rock, which caused it to rise to the surface. These massive impacts sent faults deep into the Moon's surface that made pipes for the rising lava. The mantle underneath that created "basins," which are artificial hollow places containing water, rose closer to the surface, making it a shorter path to the surface.

Lunar lavas generally erupted from narrow openings, which poured out and ponded into the lower plains. When it erupted on an inclined surface, the lava could flow downhill and even create river-like channels. On the Moon these formations are called sinuous rilles. Some may even run up to several 100 km before pouring out their lava onto lower surfaces. This process of flooding results in large, flat lava sheets that covered the basins. Lava was thicker in the center of the basin and thinner toward the edges. Since lava is heavier than the surrounding crustal rock, it compresses the bedrock underneath. The thicker areas, which would be in the middle, did this more than the thinner, less heavy lava on the sides. This made the center sink as the outer areas remained raised.

Lunar volcanoes, which are also called lunar "domes," are smooth sided with low levels of incline. This is because lunar lava has a low viscosity. Most of the lunar domes are 5–20 km across, which also have small pit craters. The Marius Hills region contains a large number of lunar domes. Some of the domes in this region are very steep sided.

Volcanism on the Moon is different in many ways from volcanism on Earth. On Earth it is an ongoing process. On the Moon, volcanism is 3–4 billion years old compared to Earth's at most 100,000 years. The oldest rock on Earth is about 3.9 billion years old, and the oldest seafloor basalts is 200 million years old. The Moon is considered a "dead planet" because no evidence shows recent volcanic or geologic activity. Earth's volcanoes occur within long linear mountain chains, also with very large, very old craters. Lunar mare is found on one side of the Moon that covers one-third of the Moon and less than 2 % of lunar areas on the two-thirds. The primary factors that control the volcanism on the Moon appear to be surface elevation and crustal thickness.

Shielded from Earthbound eyes, the far side of the Moon is home to a rare set of dormant volcanoes that changed the face of the lunar surface, a new study finds. Data and photos from NASA's LRO reveal the presence of now-dead silicate volcanoes, not the more common basaltic volcanoes that litter the Moon's surface, researchers said.

"Most of the volcanic activity on the Moon was basaltic," said Brad Jolliff of Washington University. "Finding other volcanic types is interesting, as it shows the geologic complexity and range of processes that operate on the Moon, and how the Moon's volcanism changed with time."

Because the Moon's rotation has been affected by tidal forces between Earth and the Moon, only one side of the Moon is visible from Earth. The far side of the Moon—sometimes referred to inaccurately as the "dark side"—was hidden from view until 1959, when Soviet Union's *Luna 3* spacecraft took the first photos of the region.

When NASA's Lunar Prospector probe circled the Moon in 1998, it revealed a highly reflective plain lying between two ancient impact craters. Known as the Compton-Belkovich region, this part of the Moon contains thorium and other silicate rocks, suggesting a more involved type of volcanic activity than that which created the Moon's well-known dark basaltic plains known as maria, or seas.

It wasn't until the LRO captured higher-resolution images of the region that this volcanic activity could be confirmed, however. The spacecraft found a number of domelike features with steeply sloping sides—telltale signs of lunar volcanoes.

Jolliff said that the domes likely formed by lava probably came from deep within the Moon. It flowed upward through cracks to pool just beneath the surface, where it pressed out to form large domes. Lava continued to work its way to the surface throughout the area, building other, smaller volcanic domes. Some areas then collapsed, creating the irregular depressions observed by LRO's camera, researchers said.

Scientists have known for years that volcanoes on the Moon filled in craters to form the dark maria visible from Earth's surface. However, those lava flows are basaltic in nature. "We've known for a while that the Compton-Belkovich had an unusually high thorium content," said Timothy Glotch of Stony Brook University, New York. "Now we can positively say that that thorium is related to these silicic volcano materials."

Volcanic Structures on the Moon

Glotch, working with another team, was the first to identify non-basaltic volcanoes on the near side of the Moon. Due to their highly reflective surface, this group was also originally noticed by the Lunar Prospector.

However, lava from the surrounding maria may have also partially concealed the volcanoes, so some details of the region's geologic history could have been hidden, researchers said. But the volcanoes on the far side have no maria nearby to hide their features. The complete view of the volcanism in the area lies open to examination. Similarly, they are surprisingly free from impact craters, which reveals a great deal about their age, researchers added.

The early life of the Solar System was violent, with rocks scarring the surface of the planets and their Moons. Features that lack this scarring formed after things had calmed down.

Jolliff and his team estimated the age of the Moon's rare far side silicate volcanoes to be about 800 million years old. Such an age would extend the volcanic activity of the Moon by 200 million years, they said.

According to Glotch, the discovery of non-basaltic volcanoes on the far side of the Moon "shows that the Moon is more compositionally diverse than we realized before this new age of lunar exploration. As scientists, we're still digesting all this relatively new data and working to understand what it means in terms of lunar history."

Volcanic rocks are produced by the melting of other rocks. On Earth, this can be performed in a number of ways, including the complicated method of subduction. On the Moon, it appears that volcanic rocks were produced in only one way—that of partial mantle melting. Recycling of lunar crust did not occur or it was very rare. The melts produced on the Moon are basalts.

Basalts are common products of partial mantle melting on the terrestrial planets. This is mainly due to broad similarity of their mantle compositions. For partial melting to occur on the Moon, temperatures greater than 1100 °C at depths of about 200 km are required. The bulk eruption styles appear to be in the form of lava flows. There is also widespread evidence of fire-fountaining forming pyroclastic deposits (typically glass beads).

Mare Lava Flows

The mare plains were formed by low viscosity basalt lavas erupting in large volumes. Current usage describes the lavas as maria and the basins as mare.

The best descriptions of flows come from the Apollo missions. In Mare Imbrium, lavas were erupted from the southwestern

Fɪɢ. **6.2** Lava flow on Mare Imbrium. (Credit: NASA)

edge, in three successive flows. The flows extend for 1200, 600, and 400 km, and at a slope of about 1:1000. Flow scarps bounding these flows vary from 10 to 63 m in height. Combined, these flows cover about 2×10^5 km^2 with a volume of about 4×10^4 km. The concept of very long flow distances over flat terrain is supported by leveed flow channels and localized lava ponds (dammed by wrinkle ridges) (Fig. 6.2).

Such thin flows would be expected to cool quickly. The apparently contradictory long flow distances was explained by the unusually low basalt viscosities. Supporting evidence for lava flows about 20 m thick came from petrological analysis of samples from *Apollos 11, 12,* and *15.* Also, three distinct lava flows were observed in the wall of Hadley Rille. These three flows were within 60 m of the mare surface. Crater densities vary between different flows, indicating substantial periods of time between individual eruption cycles. Crater-density ages indicate that lava production in the main mare basins lasted about 500 Ma. Infrared reflectance also indicates a large variety of compositions. Apollo has only sampled some of these varying compositions. Also, Apollo has "missed" the oldest and the newest lavas.

Radiometric dates show that the prominent basalt eruptions over the entire Moon lasted from about 3.9 to 3.1 Ga. Crater densities suggest that minor eruptions may have continued to 2 Ga. The start of maria eruption is unknown, but the oldest basalt sample is dated at 4.2 Ga. This was extracted from a highland breccia. No corresponding outcrop has been found.

Sinuous Rilles

These are meandering channels that commonly begin at craters. They end by fading into the mare surface, or into chains of elongated pits. Sizes range from a few tens of meters to 3 km in width. Lengths range up to 300 km. Channels are U-shaped or V-shaped, but fallen debris (from the walls, or crater ejecta) have generally modified their cross-sections. Most sinuous rilles are near the mare basin edges, although they are found in most mare deposits. *Apollo 15* confirmed the theory that sinuous rilles were analogous to lava channels and collapsed lava tubes (Fig. 6.3).

Lunar rilles are much larger than their terrestrial equivalents. This is thought to be due to a combination of reduced gravity, high melt temperature, low viscosity, and high extrusion rates.

Mare Domes

Domes are defined as broad, shallow landforms. These are convex and circular to oval in shape, and occur on the mare basins. Eighty low domes (2–3° slopes) have been mapped. Diameters range from 2.5 to 24 km, and heights vary from 100 to 250 m. Most of these domes are in the Marius Hills area, where they accompany other interesting geological features (possible collapsed lava tubes, etc.). Some domes have summit craters or fissures.

Observation and mapping of lava domes still continues. Due to their low profile, orbiter and Clementine images do not tend to show them very well, due to the typically high solar angle on such images (they have short shadows). Hence observation is better from Earth, where the solar angle can be deliberately chosen. A number of amateur groups, such as the British Astronomical

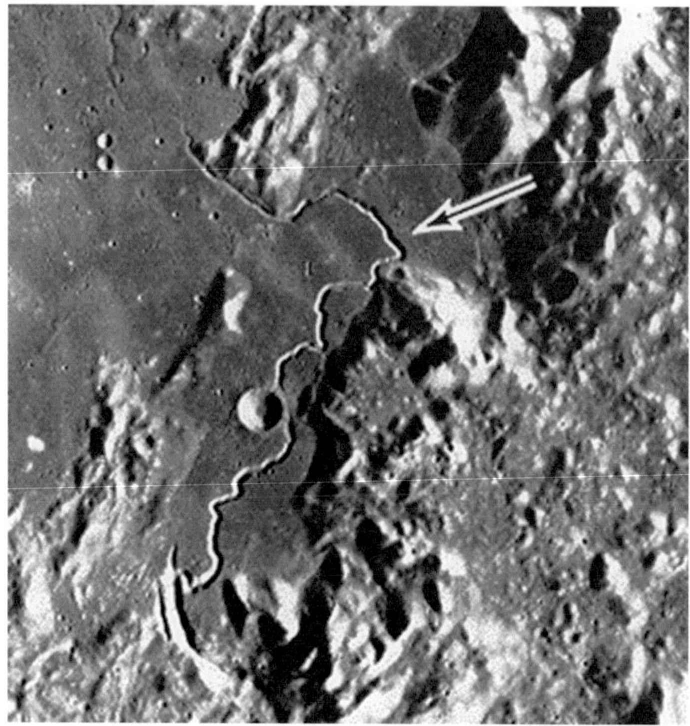

Fig. **6.3** This photo shows the Hadley Rille on the southeastern edge of Mare Imbrium. It is fairly well known because *Apollo 15* landed there. (Credit: NASA)

Association and the Association of Lunar and Planetary Observers, run active mapping and observation programs (Fig. 6.4).

Note that these should not be confused with terrestrial "volcanic domes." On Earth, the term "volcanic dome" implies a silicic, highly viscous lava (for example, Mt. St. Helens). No such lavas have been found on the Moon. Formation of lunar domes is still an unknown area. They might be due to the eruption of more viscous basalt lavas, by intrusion to form shallow laccoliths, or by mantling of large blocks of older rocks.

Lava Terraces

Small lava terraces have been observed within some craters and along mare-highland boundaries. These have been interpreted as "shorelines" left as lava has withdrawn (e. g., back into a vent or lower basin).

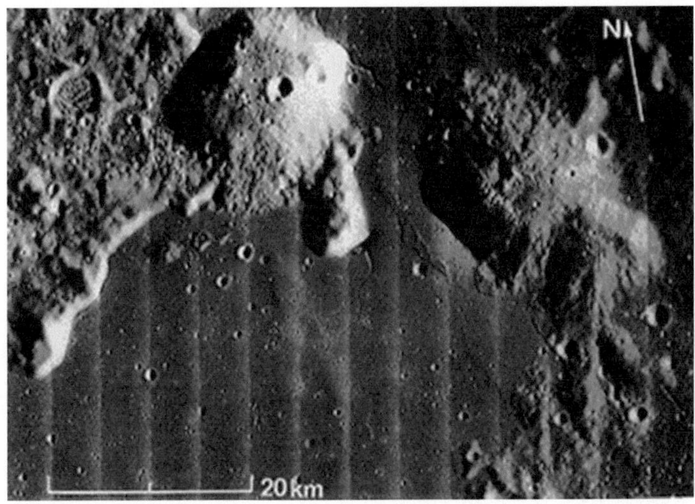

Fig. 6.4 This image shows two mare domes in northwestern Mare Imbrium. They are located on the rim of an old, mare-filled impact crater. Note that these domes differ in color both from mare basalts and from typical highland rocks. They may mark a rare instance of non-basaltic lunar volcanism

Cinder Cones

Terrestrial cinder cones are formed from lava bombs, "cinders," etc., erupted explosively from a central vent. The volume of each cinder cone is much smaller than the total basalt erupted from the cone. On the Moon, cone-like landforms have been observed in association with rilles. Most are less than 100 m high with diameters of 2–3 km, and have a low albedo. Some have summit craters (with diameters of less than 1 km). Lines of cones are thought to be fissure vents.

Dark-Haloed Craters and Pyroclastic Deposits

Extrapolation from terrestrial cinder cones suggests that lunar deposits should be broad, pancake-shaped, and low. This is mainly due to the low gravity. Although no such structures have been identified for certain, this might be an explanation for the numerous dark-haloed craters. Again, these are usually along the margins of mare basins, or along rilles and other lineaments. Most are shallow with dark rim deposits 2–10 km in diameter.

Apollo 17 managed to sample the dark-haloed crater "Shorty." This turned out to be non-volcanic, but it penetrated a loose layer of glassy and partly crystalline droplets. These were orange and black in color. They proved to be volcanic in origin and to predate Shorty. It is unknown whether other dark-haloed craters might be volcanic. Unambiguous dark-haloed impact craters can be found in highland areas adjacent to the mare. It has been suggested that these areas are due to highland-derived debris (i.e., regolith) overlying very early mare basalts at the rims.

Whether these black deposits are from the central crater, or exhumed by an impact, they are still of interest due to their relation to explosive volcanic activity. Such active eruptions imply a high volatile content in the melt. The material sampled from Shorty proved to be low-viscosity but rich in iron and titanium. These droplets tend to be mantled in condensed volatiles such as zinc and sulfur. Compositions of the droplets themselves match those predicted by partial melting of the mantle. It is still possible that the melt was contaminated by assimilation of wall rocks (which would probably be basalts).

Filling of the Mare Basins

The mare basins were formed by large impacts. These impacts would have formed large cavities that collapsed almost immediately. These collapses led to large-scale fracturing in the crust. These fractures are thought to be the conduits along which the basaltic magmas ascended to the surface. Such conduits appear to occur just within the basin edges—i.e., where large slump zones would be expected.

What caused the melting? This might be due to the thinner overlying rock (i.e., basin-filling ejecta) having a lower thermal bulk conductivity. Hence, the mantle isotherms would move upwards (i.e., a house with a thinner roof would have a warmer roof than a house with a thick roof; hence the loft would be even warmer!). This warming in the upper levels of the mantle might be sufficient for partial melting to occur. Another alternative is that the sudden removal of the overburden led to an isostatic uplift (i.e., a terrestrial mountain range lifts up as the top is eroded). This reduction in pressure could lead to the formation of partial melts. On Earth, ocean ridges form melt by releasing the confining pressure on the upwelling mantle underneath.

Infilling may have occurred for quite a while after the initial basin excavation. The Imbrium Basin is currently dated at 3.9 Ga, but the Mare Imbrium lavas are dated at 3.3 Ga (*Apollo 15*).

Most of the mare basalts are confined to basins, but occasionally they overflow into adjacent terrains. Also, many of the basins are not infilled—especially on the far side. This is thought to be due to the Moon's thicker crust on the far side. There is also a correlation with age. The youngest basins tend not to show any volcanic activity, while older basins are infilled. There are exceptions to both of these trends.

Mare Orientale, the youngest large lunar basin, is only partly filled with basaltic lavas. Its interior is dominated by a flat layer of impact melts—possibly non-basaltic in composition. Later basaltic melts have spilled over onto these impact melts. This is thought to represent the early stages of basin infilling. Eruptions begin with infilling of the innermost ring, and then along the ring's concentric fault systems. Vents and domes are also well-preserved along the next ring (the Rook Mountains), i.e., eruption vents are along ring faults.

Mare Imbrium is older. Its ejecta deposits cover a large part of the near side. No impact melt is visible. Lava flows cover most of the area between the inner ring and the outer (topographic) ring (the Appennines). The Serenitatis Basin is older still. Its interior rings are completely flooded. Researchers have modeled the evolution of this basin with three stages. They suggest that the Stage I lavas erupted in the southern basin rim. Some spilled over into Mare Tranquillitatis (sampled by *Apollo 11*).

During this time, the center of the basin sank under the load of the Stage I lavas. Stage II lavas erupted into this newly made depression. These are approximately concentric to the Stage I lavas, and clearly show fewer craters (hence are younger). Further subsidence occurred due to the extra loading. Stage III lavas occupy the very center of the basin. Some Stage III lavas managed to spill into the Imbrium Basin, illustrating the very low gradients present. This model suggests that central mare thicknesses may exceed a few km. It also accounts for the compressional ("wrinkle") ridges.

Most petrological models suggest that the basalts formed by partial melting at depths of 200–400 km. The melts are of lower density than the surrounding rocks—hence they tended to rise to the surface. In the brittle upper crust, faults and fractures would have formed the conduits. Fissure vents only have to be 10 m wide to produce the observed high eruption rates. Vertical speed of the rising melts had to be greater than 0.5–1 m/s in order to maintain lava fountains at the fissures. Vesicles (bubbles) are observed with lavas and pyroclasts. Unlike on Earth, water could not have been a significant gas component due to its scarcity. It's been proposed that CO was the main gas phase that drove the volcanic activity. Some researchers found that only 250–750 ppm CO was required to disrupt magmas at 15–40 m depths. i.e., bubbles grow so big that they burst and spatter the melt (fountaining).

So if the Moon is now volcanically inactive, then what's going on with the Ina Caldera? This 2-mile-long D-shaped patch sits atop a low, broad volcanic dome or shield volcano, where lavas once oozed from the Moon's crust. Blobs of older, crater-pitted lunar crust (darker blobs) rise some 250 ft. above the younger, rubbly surface like melted cheese on pizza. Brighter areas on the Moon generally indicate younger surface features. Solar and cosmic radiation darkens airless worlds like the Moon and asteroids over the long haul of time. Not only is the bumpy region lighter-toned, but it appears to have far fewer craters, another sign of its relative youth.

Volcanic Activity Alive and Kicking...Barely

The Moon, thought to be cold and dead, is still alive and kicking—barely. Scientists have found evidence for dozens of burps of volcanic activity, all within the past 100 million years—a mere blip on the geologic timescale. And they think that future eruptions are likely, although probably not within a human lifetime.

For a world thought to have gone cold long ago, the discovery points to a place that still releases internal heat in fits and starts, says Mark Robinson, a planetary scientist at Arizona State University, Tempe, and a co-author on the new study. "The big story is that the Moon is warmer than we thought," he says.

Much of the Moon's near side is paved over in dark plains of basalt called maria. The activity that created these ancient lava fields peaked about 3 billion years ago and petered out 1 billion years ago. But a strange geologic structure called Ina has intrigued scientists for decades, and in 2006 they found evidence that it was indeed a volcanic vent that was active as recently as 10 million years ago. But Ina remained the anomaly.

Scientists working with NASA's LRO, which has been orbiting the Moon since 2009, decided to look more closely for possible volcanic structures. The orbiter's camera can capture images with resolutions of between 50 and 200 cm per pixel—the sharpest pictures of the lunar surface ever taken.

On the near side of the Moon they found 70 structures that appear to be lava flows, ranging in size from 100 to 5000 m across. The researchers think that some of these features are the remnants of low-lying shield volcanoes, which ooze out a soupy lava.

'Blobs' would be an apt term for them, says Sarah Braden, who recently completed her doctorate at Arizona State and was lead author of the study, which appears online in *Nature Geoscience*. The team says that the flows must be relatively new because of the sharp contacts between the lava and the underlying rock—distinct boundary lines that moonquakes and micrometeorite impacts would have eroded over time.

The researchers estimated ages for the largest of the flows by counting craters on them, a technique based on the idea that a young surface will be less heavily pocked with craters than an old one. Braden says the youngest flow resides in a tub-shaped depression near the western edge of the Sea of Tranquility and is just 18 million years old. And if there are flows that young, Robinson says, eruptions are apt to happen again—although not necessarily in the near future. "I doubt if I'll live long enough to see it happen," he says.

Mark Wieczorek, a planetary scientist at the Institute of Earth Physics in Paris, says the result shows that the Moon has not cooled as quickly as scientists once thought. He says it makes sense that the structures were discovered on the Moon's near side, which is heavily enriched in radioactive elements that produce heat and help spur volcanic activity. To this day, the deep near-side crust is a couple of 100° warmer than its counterpart on the far side.

And so the LRO team may have detected one of the Moon's "last gasps" of volcanic activity before it dies completely, Wieczorek says. "The body is still warm. From time to time, you'll still see a few twitches."

Whole Lot of Shaking Going On

Many people think of the Moon as a dead planetary body, in terms of seismic activity. Quite the contrary. When astronauts return to the Moon, they may need quake-proof housing.

That's the surprising conclusion of Clive R. Neal, associate professor of civil engineering and geological sciences at the University of Notre Dame, after he and a team of 15 other planetary scientists reexamined Apollo data from the 1970s. "The Moon is seismically active," he told a gathering of scientists at NASA's Lunar Exploration Analysis Group (LEAG).

A moonquake is the lunar equivalent of an earthquake. They were first discovered by the Apollo astronauts. Moonquakes are much weaker than the largest earthquakes, though they can last for up to an hour, due to the lack of water to dampen seismic vibrations.

Information about moonquakes came from the Apollo Passive Seismic Experiment (APSE), a network of four seismometers placed on the Moon by Apollo astronauts from 1969 to 1972. The instruments placed by the *Apollo 12, 14, 15* and *16* functioned perfectly until they were switched off in 1977 due to cost-saving measures (Fig. 6.5).

While the network was operational, it clearly demonstrated that the Moon is seismically active, albeit on a smaller scale than Earth. The Moon is less active because it is a "one plate planet" and does not, therefore, contain plate boundaries, the sites of large seismic events here on Earth. But plate boundaries aside, the Moon exhibits seismic activity on a similar scale to that of intraplate (interior of a tectonic plate) settings on Earth.

Over the last three decades, many researchers, most notably perhaps Yosio Nakamura of the University of Texas in Austin, have mined the Apollo seismic database and identified four different types of lunar seismic events.

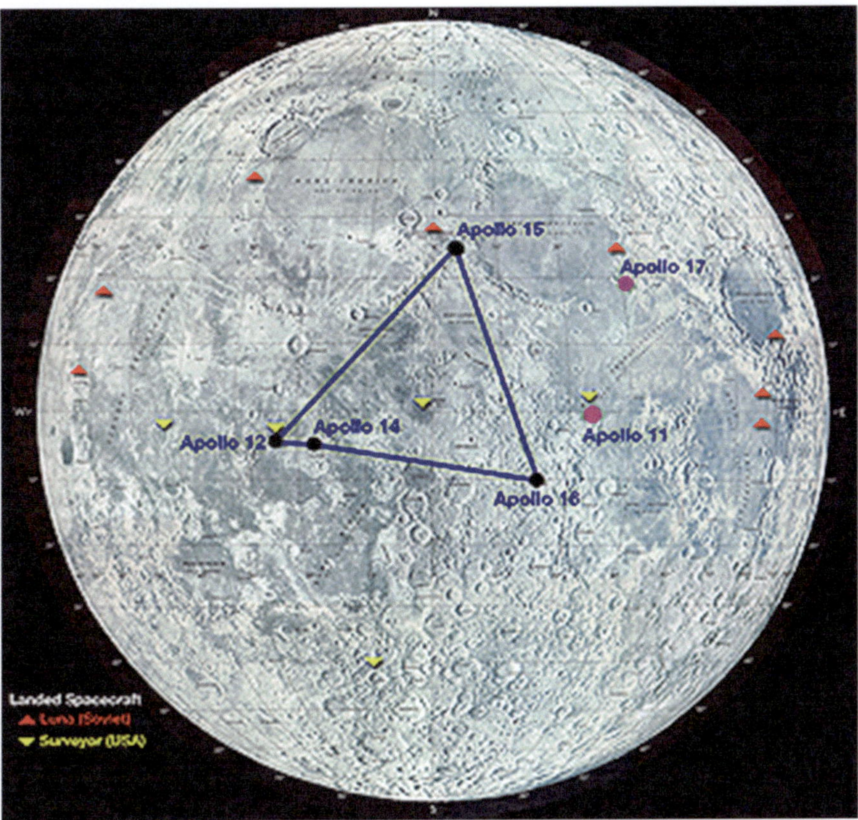

Fig. 6.5 The Apollo Passive Seismic Experiment involved deployment of a network of seismometers (outlined in the large triangle) on the near side of the Moon. The seismometer at *Apollo 11*'s landing site failed after 21 days. The *Apollo 17* seismometer was intended to determine the shallow structure of the site and was switched off once the experiment was complete. (Credit: NASA)

- *Deep Moonquakes.* These occur up to 700 km below the Moon's surface, resulting from tidal stresses caused by the gravitational tug of war between Earth, the Moon, and the Sun. Deep moonquakes are also low in energy but are the most abundant type, with more than 7000 events recorded during the 8 years of monitoring. These are small-magnitude events (generally less than magnitude 2) occurring approximately halfway between the surface and the center of the Moon. Deep moonquakes originate from specific locations, or "nests," that produce a quake of unique waveform. To date, researchers have identified 318 nests from the Apollo dataset.

- *Meteorite Impact Vibrations.* These are rumbles or vibrations of the surface when a meteoroid slams into the surface. Although most of the energy of an impact is expended in excavating a crater, some is converted to seismic energy. Between 1969 and 1977, the APSE network recorded more than 1700 events representing meteoroid masses between 0.1 and 1000 kg. Events generated by smaller impacts were too numerous to be counted.

- *Thermal Moonquakes.* These are small and were recorded in several locations by the Apollo seismometers. They are the result of daily temperature changes that produce stress and strain of lunar rocks in young craters, eventually causing them to crack, or "slump," as evidenced by the repetition of signals with nearly identical waveforms at the same time during each lunar day. Seismic activity begins abruptly 2 days after lunar sunrise and rapidly deteriorates after lunar sunset. Of the four types of moonquakes, these events emit the lowest energy.

- *Shallow Moonquakes.* These are the strongest type of lunar seismic events. Such quakes were originally referred to as "high frequency teleseismic events," and seven of the 28 events that the APSE network recorded were greater than magnitude 5. The depths at which these moonquakes originate are "shallow," likely between 50 and 200 km, although the exact depths are unknown because all recorded events were outside the perimeter of the APSE network. These events are relatively rare and not correlated with tides. Although they appear to be associated with boundaries between dissimilar surface features, the exact origin of these events is still unclear. A recent development by Nakamura and colleagues suggests shallow moonquakes may originate from interaction of the

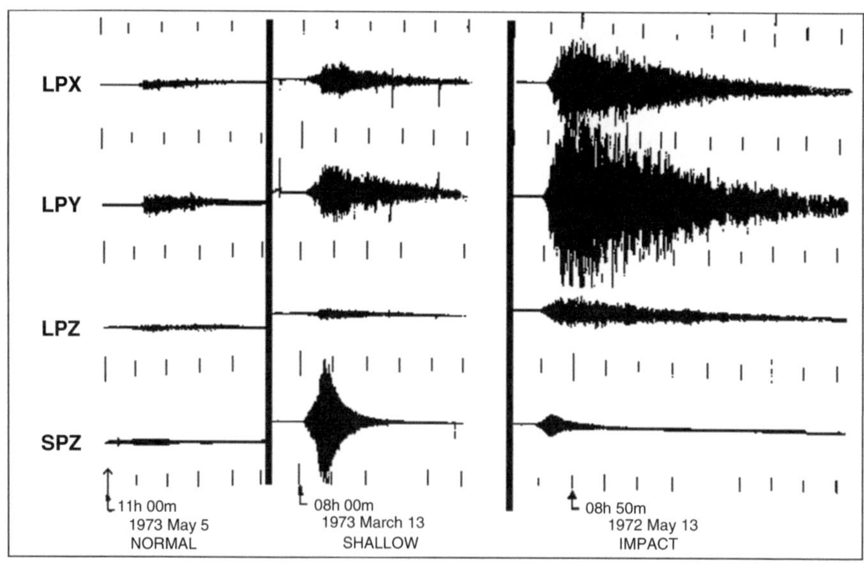

Fig. 6.6 Three types of moonquakes are shown here. The top three rows are the recordings of a three-component long-period seismometer; the bottom row are the registrations of a short-period instrument

Moon with nuggets of high energy particles ("strange quark matter") originating from a fixed source outside the Solar System (Fig. 6.6).

The first three kinds of moonquakes mentioned above tend to be mild; however, shallow moonquakes can register up to 5.5 on the Richter scale. On Earth, quakes of magnitude 4.5 and above can cause damage to buildings and other rigid structures. Furthermore, shallow moonquakes lasted a remarkably long time. Once they got going, all continued more than 10 min. "The Moon was ringing like a bell," Neal says.

On Earth, vibrations from quakes usually die away in only half a minute. The reason has to do with chemical weathering, Neal explains: "Water weakens stone, expanding the structure of different minerals. When energy propagates across such a compressible structure, it acts like a foam sponge—it deadens the vibrations." Even the biggest earthquakes stop shaking in less than 2 min.

The Moon, however, is dry, cool, and mostly rigid, like a chunk of stone or iron. So moonquakes set it vibrating like a tuning fork. Even if a moonquake isn't intense, "it just keeps going and going," Neal says. And for a lunar habitat, that persistence could be more significant than a moonquake's magnitude.

"Any habitat would have to be built of materials that are somewhat flexible," so no air-leaking cracks would develop. "We'd also need to know the fatigue threshold of building materials," that is, how much repeated bending and shaking they could withstand.

The shallow moonquakes and those caused by meteoroid impacts pose potential (and very real) hazards to establishing a long-term habitable facility on the Moon. Still, it is important to note that the seismicity generated by meteoroids is unlikely to pose a significant threat to any structure, unless the impactor's mass is on the order of tons and the impact is close to the site of the Moon base. Shallow moonquakes, however, contain more energy at high frequencies than quakes on Earth of similar magnitudes.

Of critical importance to a lunar base is the amount of ground motion associated with shallow moonquakes. As already noted, seven of the 28 shallow seismic events recorded by the APSE network had magnitudes of five or more. On Earth, such quakes can cause structural damage around the epicenter, often manifested by the propagation of cracks through and along walls. In a lunar setting, a magnitude five or greater quake could breach the seal of a Moon base, resulting in a catastrophic loss of oxygen, which would devastate the habitat.

Currently, we do not know the cause or exact locations (especially the depths) of these strong, shallow moonquakes, but they are in many respects comparable to infrequent, highly destructive intraplate earthquakes. As a first order approach, researchers have modeled ground motion and acceleration using equations derived for terrestrial quakes (there are no such equations as yet for the lunar scenario). Applying such terrestrial models to the lunar environment is, at present, the best estimate we can make. However, there are distinct differences in terms of seismic wave transmission between the Moon and Earth that could render the above estimates totally inadequate.

For example, researchers have observed the maximum signal from a shallow moonquake to last up to 10 min, with a slow tailing off that can continue for hours in total duration, demonstrating that signals dampen less while traveling through the Moon than they do traveling through Earth. This effect suggests the mechanical properties of lunar rocks are distinct.

Studies have demonstrated the dramatic damping effect that water has on seismic energy. Thus, seismic energy is more efficiently propagated through the Moon, which is incredibly dry. Significantly, moonquakes tend to produce seismic waves of higher frequency than earthquakes. This consideration is important for wave transmission through the lunar interior as well as interaction with the near surface.

The material, or "regolith," on the lunar surface has formed through micro- to macro-sized meteoroid impacts. This incredibly dry unconsolidated material tends to scatter seismic waves. Modeling of this scattering using frequency-dependent diffusion appears to allow a statistical approach to quantifying the data recorded by the APSE network. Using this model, earlier researchers highlighted the fact that there is very low attenuation of waves in the surface zone.

Although the APSE network has defined two potential hazards for any permanent habitation on the Moon, accurate risk assessment is difficult to formulate because of the limitations of the current lunar seismic database. By integrating the existing seismic data with strategically acquired new data, researchers can not only better understand potential hazards but also can better refine scientific models about the Moon's interior structure and formation.

The small area covered by the APSE network has resulted in difficulty in estimating epicenter locations for seismic events. Although attempts have been made to locate the origins of both deep and shallow moonquakes, exact locations require knowledge of the composition of the lunar mantle. The small area extent of the APSE network has resulted in reduced resolution of seismic information from deep in the Moon's interior, limiting interpretation.

For example, seismic data have been interpreted to indicate the presence of garnet in the lunar mantle. However, the same seismic data were also interpreted to represent an increased proportion of magnesium-rich olivine—a conflicting view. The presence of garnet in the lunar mantle is also supported by the trace element ratios of some lunar volcanic glasses. Resolving these differences to evaluate the Moon's composition is important to any seismic risk evaluation, as it affects how seismic waves travel through the body. More seismic data are necessary to better understand the composition.

Likewise, very little seismic information is available regarding the nature of the lunar core, as the APSE network was located on the near side of the Moon. A large meteoroid impact on the far side of the Moon produced a seismic wave that was significantly delayed in reaching one of the Apollo seismic stations. This event suggested the presence of a molten core approximately 720 km in diameter, but no additional data were collected to confirm this.

Geochemical and magnetic data also suggest that the Moon does indeed have a small core with a diameter estimated to range from approximately 500 to more than 800 km. However, the composition of this core has been reported to be anything from iron to iron sulfide to ilmenite (which also contains iron). These compositions have important implications for the thermal history of the Moon. For example, if the core is predominantly iron, it would be solid, but if it is iron sulfide, it could still be (partially) liquid. Appropriate seismic data could rigorously test the nature and composition of the lunar core.

The location of the APSE also did not provide much information regarding deep seismic activity on the lunar far side.

Early research reported 109 distinct nests of deep moonquakes, but only one of these was located on the far side. A reevaluation of the APSE database found two more indisputable far side nests with several other nests having epicenter locations on the near-side, but uncertainty remains. These far side nests seem to produce seismic waves that stop while traveling through the Moon's interior, suggesting there is a "plastic," or partially molten, zone located somewhere in the deep lunar interior. Two major questions remain: (1) Are there other nests on the lunar far side that were not detected by the APSE? and (2) What is the nature and extent of the purported plastic zone.

More seismic data will not only help resolve these questions about the Moon's interior but also could inform hypotheses about the Moon's formation. The lunar magma ocean model has become the cornerstone of our understanding of Moon evolution. This model postulates that immediately after formation, the Moon underwent a global melting event that affected approximately the outer 500 km of the Moon, or even all of it.

The APSE data have been interpreted to indicate the presence of a seismic discontinuity (a surface at which velocities of seismic waves change abruptly) at approximately 400–600 km beneath the equatorial Moon that could represent the base of the lunar magma ocean, but it is unclear whether this extends around the whole Moon. If it does, then there is evidence for the global melting event, although it may indicate that only the outer "skin" of the Moon melted. If it does not, then a rethinking of the magma ocean model is required. Again, more seismic data are necessary.

Predicting where shallow moonquakes will occur is of prime importance for the next phase of lunar exploration. Current data make this difficult because only 28 such events were recorded before the APSE was shut down.

The relatively small number of events and the error in locating epicenters has limited the statistical significance of relating these shallow seismic events to tectonic features or a possible extra-Solar System origin. Thus, the causes and locations of shallow moonquakes are not known with any high degree of confidence. However, the existing data indicate that one magnitude five or greater event occurs each year. Because the shallow quakes are potentially the most destructive, such knowledge must be obtained in the near, if not immediate, future.

Thus, an urgent need exists for a long-lived global seismic network to be established on the Moon. Over the last three years, without the help of astronauts, an international team has been investigating the intricacies of remotely deploying the Lunar Seismic Network (the "LuSeN mission").

The seismometer that was chosen for this investigation was the Centre National d'Etudes Spatiales' "very broadband seismometer" that was developed for deployment on Mars, and is more sensitive than those used in the APSE. However, the outcomes of this investigation highlighted major technology gaps in the remote deployment of seismic networks on airless bodies. The major issues include developing long-lived power supplies (with a 6-year minimum lifespan) that can survive the long lunar night and planning for a hard versus soft landing deployment of the sensors.

The seismometers that formed the APSE network were powered by RTGs (Radio-isotope Thermoelectric Generators) with a design output of 73 W. The LuSeN team has estimated that the maximum power required for the very broadband seismometer, integrated with computerized recording and transmission components, is between 2 and 4 W. It is unclear whether electrical power in this range can be maintained for a minimum of 6 years using conventional solar-battery technology. An Apollo-sized RTG for each seismometer package, however, would be overkill.

What is required is the development of so-called "mini-RPS" (Radioisotope Power System) units that supply between 0.1 and 10 W of power. Although this concept has been explored by NASA, actual fabrication and testing of such units is not occurring, nor is it planned to occur in the near future. However, the concepts presented by researchers at NASA's Jet Propulsion Laboratory in Pasadena, CA, in 2004 are tantalizing for the LuSeN mission and any mission that attempts to deploy a global seismic network on an airless body. For example, the mass of the RPS unit would be between 3 and 6 kg, resulting in a total mass for each seismometer package of less than 8 kg—much lighter than using conventional solar-battery-power supplies, which cause the packages to weigh much more than 10 kg each.

In general, mass is a critical limiting factor. If the mission calls for a "hard landing," each seismometer package will need cushioning material to protect the delicate instrumentation as it drops from a low-altitude orbiter. But if a "soft landing" deployment is planned, each package will need its own individual "retro-rocket assembly" to slow its landing. Either option adds more mass to each package, and, especially if conventional solar battery power technology is used, a network cannot be established with a single launch. At least two launches will be required to establish a minimum network of eight seismometers. The mission will, therefore, be cost-prohibitive using the current mission programs available through NASA.

At this time it is suspected, but not known, that seismic events could seriously compromise a permanent lunar habitat. To fully evaluate this risk, a long-term, global lunar seismic network needs to be established. Now is the time to start developing the new power-generation systems so that the lunar seismic network can be established to monitor and locate seismic hazards before we humans can make the Moon our home.

New Computers Uncover Old Moonquakes

In the category of "good research takes time," scientists have just discovered more than 100 quakes that occurred on the Moon three decades ago.

Seemingly combining archaeology and modern computer science, several groups of researchers have exhumed data from the Apollo landings and taken a closer look at them to improve their understanding of what the Moon is made of and why its inside trembles.

"It's a very exciting dataset," said Dr. Yosio Nakamura, a geophysicist at the University of Texas, who was among the scientists who analyzed the seismographs in the 1970s and continues to work on them today.

The lunar landers for *Apollo 11–16* carried seismometers, the same type of instruments that measure earthquakes on Earth, designed to be left behind and to continue to radio back data after the astronauts left. *Apollo 11* left a prototype that failed after a month, and *Apollo 13*, which was almost lost when its oxygen tank exploded, never made it to the lunar surface. But the four other seismometers, from *Apollo 12, 14, 15,* and *16,* recorded some 12,500 seismic events through 1977, when NASA turned the network off.

As expected, the Moon shuddered when hit by small meteors. There were a handful of deliberate tremors when discarded parts of the Apollo spacecraft were sent crashing onto the lunar surface.

The seismometers also recorded 28 shallow "moonquakes" over 8 years, the largest about magnitude 5.0. But more unexpected and still not well understood were numerous tiny moonquakes, several a day on average, that occurred far below, 500–750 miles down, about halfway to the Moon's center.

"That's deeper than any earthquake we see on the Earth," said Dr. Catherine L. Johnson, a geophysicist at the Scripps Institution of Oceanography in San Diego. "It tells you something about the ability of the middle of the Moon to fracture."

Nearly identical squiggles emerged from the Moon seismographs over and over, indicating that certain parts of the Moon's interior broke repeatedly in the same way, almost like someone cracking his knuckles. Dr. Nakamura and other scientists counted 108 of these regions, which they called nests. The rate of moonquakes seemed to ebb and flow over every 27 days, the time it takes the Moon to circle Earth, suggesting that they were caused by the pull of tidal forces.

Almost all of the deep moonquakes originated on the near side of the Moon, the side that always faces Earth. This meant that either the structure of the far side is different and does not experience moonquakes or moonquakes do occur, but the waves dissipate when they hit a still molten core.

That was the unsatisfying, unresolved conclusion to Dr. Nakamura's initial studies 20 years ago. He also could not figure out the sources for more than 9000 of 12,500 events in the seismographs at that time.

The difficulty then lay in the limits of early computers. The four seismometers on the Moon were advanced for their era, recording the data digitally and radioing the information to Earth in real time. But mainframe computers were incapable of processing that much data.

Instead, Dr. Nakamura and his colleagues took the computer-friendly digital data and printed them on old-fashioned rolls of paper, then placed the paper on light tables and identified moonquakes by eye. "We still have those pieces of paper," he said.

Now he has reanalyzed the 9000 unidentified events using computers. In a paper published in *The Journal of Geophysical Research*, Dr. Nakamura reported that 5885 of them turned out to be deep moonquakes and that he had located about 250 new nests. Only a few of the nests were located on the far side, leaving the question about the core and far side earthquakes still unanswered.

Meanwhile, Dr. Johnson and a graduate student, Renee C. Bulow, also sifted through the original digital seismic data, looking for the signatures of small moonquakes buried in the noise that may have escaped the attention of scientists looking at the paper printouts.

The largest of the moonquake nests in Dr. Nakamura's catalog had 323 events. In their analysis of that nest, Johnson and Bulow discovered an additional 101 moonquakes. Johnson estimated that when the entire analysis was complete, perhaps 1500 new moonquakes would be added. The findings were reported at a meeting of the American Geophysical Union.

Another researcher, Dr. Amir Khan, a geophysicist at the University of Copenhagen, has also used modern analysis techniques on the moonquakes to conclude that the Moon's crust is about 25 miles thick, 25 % thinner than earlier believed. Based on the speed of the seismic waves, Khan concluded that the Moon's interior was considerably different from Earth's upper mantle, with higher concentrations of aluminum and silicon and lower amounts of magnesium and iron.

That would fit current ideas that the Moon did not form at the same time as Earth. If that had been the case, the two bodies would be expected to have a similar chemical makeup.

But even with computers wringing new answers from old data, the best hope for learning more about the Moon would be a new network of seismometers, especially on the far side.

Repeating and Swarming Moonquakes

Moonquakes can be divided into two categories: (1) repeating moonquakes with fixed source locations, and (2) moonquake swarms with variable sources locations. With few exceptions, the repeating moonquakes occur at monthly intervals near times of perigee and also show correlations with longer-term (7-month) lunar gravity variations. They originate at not less than ten different locations. However, a single focal zone accounts for 80 % of the total seismic energy detected. The epicenter of the active zone has been tentatively located at a point 600 km SSW from the *Apollo 12* and *14* stations. The focus is approximately 800 km deep. Each focal zone must be small (less than 10 km in linear dimension). Changes in record character that would employ migration of the focal mechanism have not been observed in the records over a period of 20 months. Cumulative strain at each location is inferred. Thus, the moonquakes appear to be releasing internal strains of unknown origin, the release being triggered by tidal stresses. The occurrence of moonquakes at great depths implies that (1) the lunar interior at these depths is rigid enough to support appreciable stress, and (2) that maximum stress difference occurs at these depths. If the strain released as seismic energy is of thermal origin, this places strong constraints on an acceptable thermal model for the deep lunar interior.

Episodes of frequent small moonquakes, called moonquake swarms, have been observed. One of the most significant discoveries of lunar seismology to date is the observation of moonquake swarms. Each swarm is a distinctive sequence of moonquakes closely grouped in space and time, generally containing no conspicuous event. A moonquake swarm is characterized by an abrupt beginning and ending of activity. Events are recorded at a nearly constant rate of 89–120 per day during the swarm, as compared to 1–2 per day between periods of swarm activity. During periods of swarm activity, events occur as frequently as one event every 2 h during intervals lasting several days. The occurrence as moonquake swarms shows a semi-monthly periodicity; hence they also appear to be induced by lunar tides.

The source of moonquake swarms have not been determined, but they appear to be widely distributed. Moonquakes in one swarm show a gradual increase in magnitude over a period of 2.5 days, ending abruptly with the largest moonquake yet recorded. This moonquake was distinguished not only by its size but by the high frequency of the signal it generated. The epicenter was located in the vicinity of Mare Crisium. This pattern of activity is similar to that observed from terrestrial volcanic eruptions. Thus, while the main episode of volcanism ended on the Moon 3 billion years ago, it seems there are still eruptions.

The average rate of seismic energy release within the Moon is far below that of Earth. Thus, internal convective motion leading to significant lunar tectonics appears to be absent. Presently, the outer shell of the Moon appears to be relatively cold, rigid, and tectonically stable, compared to Earth. Moonquake swarms may be generated within the outer shell of the Moon as a result of continuing minor adjustments to crustal stresses.

Is the Moon Still Tectonically Active?

The Moon's crust has been active far more recently than previously believed, according to new research. These findings raise questions about how the Moon formed and evolved.

Although Earth's crust is still shifting, driven by the churning semi-molten rock underneath it, researchers had thought that the Moon had cooled off too long ago to still have any such tectonic activity. For instance, the youngest known tectonic features on the lunar landscape until now—small cliffs in the lunar highlands resulting from wrinkling of the surface as the Moon's interior cooled and shrunk—are thought to be less than 1 billion years old, although by exactly how much is uncertain.

Now, images collected by the LRO hints the Moon has probably seen tectonic activity within the last 50 million years. In these photos, researchers spotted a dozen or so narrow, trench-like features known as graben in the lunar highlands and in the dark plains of volcanic rock known as the mare basalts. Graben are essentially troughs with two faults or cracks in the surface on either side of them. They are thought to have formed as the lunar crust was stretched.

"Overall on the Moon, you have this contracting, shrinking environment, but in some places, apparently there's this stretching extension of the crust," said study lead author Thomas Watters, a planetary scientist at the Smithsonian Institution in Washington.

The graben the scientists detected, which reach up to about 1640 ft (500 m) wide and 1.1 miles (1.8 km) long, appear relatively pristine. This suggests they formed recently. Otherwise, they would be marred more often by craters from meteor impacts over time.

"We think they're less than 50 million years old, but they could be 10 million years old, could be 1 million years old, could have happened 40 years ago," said Watters. "The intriguing picture that's emerging of the Moon is that there is recent geological activity going on."

Moonquakes detected by seismic sensors installed during the Apollo missions support the notion of recent activity on the Moon, researchers added. All in all, the Moon's interior may still be hot. "The Moon may not only have been tectonically active recently, but may still be tectonically active today," Watters said.

Models of how the Moon cooled over time suggest it was totally molten after its formation, and that it should now be contracting as it cools, forcing the surface to wrinkle. However, if this was true, such compression would have suppressed the creation of graben. These ditches, not crinkles, typically form when the crust stretches.

Instead, these findings suggest that the Moon was not completely molten after it was formed. If this were the case, the Moon would not contract strongly enough to suppress the emergence of graben.

The latest announcement is another sign that the Moon is not geologically dead. Earlier research using LRO's sharp-eyed camera found cliffs and staircase-like terrain formed by thrust faults, leading scientists to conclude that the Moon is still cooling billions of years after it coalesced.

High-resolution images from the LRO Camera, or LROC, reveal long, narrow trenches that appear to be flanked by two parallel fault systems. The troughs, known as graben, form when the lunar crust is pulled apart, fracture, and drops down, according to a NASA statement.

This discovery follows a revelation announced in August 2010 that the Moon shrunk in its recent past, and may still be contracting today. Scientists say the graben features are evidence that forces tugging the lunar crust apart overcame the forces shrinking the Moon (Fig. 6.7).

"We think the Moon is in a general state of global contraction because of cooling of a still hot interior," said Thomas Watters, lead author of the paper and a researcher at the Center for Earth and Planetary Studies at the Smithsonian's National Air and Space

FIG. 6.7 Rima Ariadaeus on the Moon is thought to be a graben. The lack of erosion on the Moon makes its structure with two parallel faults and the sunken block in between particularly obvious. (Credit: NASA)

Museum in Washington, D. C. "The graben tell us forces acting to shrink the Moon were overcome in places by forces acting to pull it apart. This means the contractional forces shrinking the Moon cannot be large, or the small graben might never form."

Scientists believe the graben formed less than 50 million years ago. The graben provide evidence that not only is the Moon shrinking, but there may still be other geologic forces at work on the surface to pull apart the lunar crust.

"This pulling apart tells us the Moon is still active," said Richard Vondrak, LRO project scientist. "LRO gives us a detailed look at that process."

The thought of lunar shrinkage is not new; scientists have long predicted the Moon contracting, but LRO's findings indicate there could still be some leftover dynamism inside the Moon driving geologic activity at the surface. If it's not still going on today, scientists know it only stopped in the recent past.

Mark Robinson, LROC principal investigator from Arizona State University, said only about half of the Moon's surface has been covered by the orbiter's high-resolution camera, which can spot features as small as 1 m, or about 3.3 ft.

"It was a big surprise when I spotted graben in the far side highlands," Robinson said. "I immediately targeted the area for high-resolution stereo images so we could create a three-dimensional view of the graben. It's exciting when you discover something totally unexpected, and only about half the lunar surface has been imaged in high resolution. There is much more of the Moon to be explored."

"Currently, a popular idea for how the Moon formed is that it was completely molten in the beginning. After a Mars-size object hit Earth very early in its history, the debris clouds from the surviving material formed the Moon," Watters said. "This may lend support to alternative scenarios that the Moon was not completely molten when it formed, that only part of it was, forming a magma ocean."

The scientists say the contractions of the Moon may also be a clue that the Moon did not form the way the planets did. That suggests that the Moon did not completely melt in the early stages of its formation. Instead, "observations support an alternative view that only the Moon's exterior initially melted, forming an ocean of molten rock."

Future research can look for more graben in LRO photos once the satellite finishes imaging the Moon, Watters said. He and his colleagues detailed their findings in the journal *Nature Geoscience*.

Even though the Moon's far side is more heavily cratered, many of the new graben discoveries lie on the lunar far side. The authors of the current study point out that the pristine condition of the graben in the LRO images is a strong indicator of their relative youth. In addition, the graben are relatively shallow, with some as little as 1 m in depth. Older features are typically deeper, since old, shallow rifts would be erased over time as they fill with debris.

In addition, some of the newly discovered graben cut across craters or distort the edges of nearby craters, showing them to be more recent than the impact. Other, smaller features are in mare regions, again revealing that they postdated the lava flows in those basins.

The researchers rule out formation of graben by impact stresses, since there are no recent craters in the vicinity. This means the young channels must be formed by tectonic processes. Comparison to similar features shows them to be consistent with the contraction of the Moon, which could come about as its interior cools and shrinks. Measurement of shallow moonquakes by instruments used during the Apollo missions is also consistent with seismic activity associated with the graben.

The Moon's interior certainly was warmer in the past than it is today, according to any reasonable model of its formation and evolution. But it's harder to know exactly how molten it was, and thus how its interior is differentiated and what has changed. The authors of the current study argue that recent graben formation is incompatible with a fully molten interior during early times, since that would create more compressional stresses rather than the stretching that forms valleys like these. However, volcanic activity is also associated with this sort of stressing, and there is no evidence of any volcanism in the eras when the graben must have formed.

Whatever the implications for models of the lunar interior, the relatively pristine graben certainly indicate the Moon is not as seismically quiet as it seems. With detailed observations from the LRO and other missions such as GRAIL (Gravity Recovery and Interior Laboratory), it will be possible to pin down exactly how the Moon continues to change in time, and refine our understanding of both the surface and interior of Earth's satellite.

Something Strange Triggering Moonquakes

Something strange, from outside our Solar System, may be crashing into the Moon and triggering "moonquakes." A new analysis of seismic data from the Moon suggests that exotic, theoretical particles called "strange quark matter" might sometimes be to blame.

Our seemingly quiet, crater-pocked satellite is actually rattled by four types of seismic activity: deep, shallow, and thermal moonquakes, and meteorite impacts. These quakes can shake the Moon's surface for an hour or more at magnitudes up to five on the Richter scale. On Earth such a quake would be strong enough to move heavy furniture and crack walls.

Each type of moonquake has a distinct seismic fingerprint, and scientists have linked each to specific causes—except shallow moonquakes. Gravitational and tidal forces cause deep moonquakes, temperature changes induce thermal quakes as the Moon moves between sunlight and darkness, and meteorite impacts, well, the name says it all. But the cause of shallow moonquakes remains a mystery.

Cliff Frohlich and Yosio Nakamura, both of the Institute for Geophysics at the University of Texas at Austin, may have discovered a clue. While analyzing data from seismometers placed on the Moon by Apollo astronauts, they noticed something intriguing. Of the 28 shallow moonquakes recorded in the data, 23 occurred while that area of the Moon faced the constellations Leo and Cancer. "You might well ask, if the moonquakes occur when the Moon is pointed in a certain direction, what's this all about?" Frohlich said.

Their answer? Something coming from outside our Solar System could be striking the Moon and causing the quakes. In a paper published in the journal *Icarus*, they suggest that strange quark matter, extremely dense particles with extra-Solar System origins, could perhaps solve the mystery.

However, strange quark matter is rather mysterious itself. Physicist Edward Witten first proposed the existence of these extremely small, dense, and fast particles in 1984. In theory, they would zip through space at hundreds of km per second. And because they would pack three types of quarks into their structure—called "up," "down," and "strange"—their density would be more than a trillion times greater than the familiar matter around us, which incorporates only two quark varieties.

So despite being as tiny as an atom's nucleus, a particle of strange quark matter could weigh several kg. "A ton of it would be the size of a blood cell," said Vigdor Teplitz of the Astrophysics Science Division at Goddard Space Flight Center.

Amazing stuff—if it exists. Strange quark matter has never been observed.

"This might be the first actual evidence that strange quark matter exists," Frohlich said. But, he poetically cautions, "there's many a slip between theory and the lip."

At the theoretical density and speed of strange quark matter, it could fly right through the Moon, like a BB shot through butter. This might leave no seismic signature at all, or it might leave seismic data on both sides of the Moon. Neither of these scenarios matches what the scientists saw.

They speculate that, instead, these little nuggets of matter may have a slightly lower density and larger girth, thus penetrating into the Moon but not all the way through, more like shooting an M&M, broadside, at a stick of butter. This revised picture better explains why their seismic data show shallow moonquake impacts only on one side of the Moon. But with these adjustments, "it's not exactly the same particle," Nakamura said. "It may be something different, something we have never thought about."

For now, the idea that strange quark matter might cause shallow moonquakes remains merely an intriguing possibility. "It's a fascinating idea," said Peter Shearer of the Institute of Geophysics and Planetary Physics at the University of California at San Diego. "[But] when you have really important new claims, you have to have a fairly high standard of proof," Shearer said.

Validating or refuting strange quark matter's contentious relationship with the Moon calls for accumulating more data—a tricky proposition since the four seismometers on the Moon shut down in 1977. Researchers are limited to this pool of old data when fishing for answers. Putting more seismometers on the Moon "would be a great idea, but it's going to be a lot of money," Shearer said.

However, shouldn't these particles, if they exist, sometimes pass through Earth as well? Some scientists think that nuggets of strange quark matter might leave a wisp of seismic data as they pierce Earth. But finding such a signal on the seismically noisy Earth would be hard; so far, combing records of earthquakes hasn't turned up any evidence. "The Moon is a better place to look for unusual phenomenon," Frohlich said. "It's quiet. Real quiet."

Until the next moonquake strikes.

Lunar Rock n' Roll

You would be forgiven for thinking that this solitary boulder on the Moon had recently rolled down a sloping wall, leaving a track behind it (Fig. 6.8).

However, NASA's Lunar Reconnaissance Orbiter Camera (LROC) is able to zoom in so closely that it is possible to see craters along the route of the 9-m boulder, which is thought to have bounced off the rim of the Schiller Crater. The tracks are in almost perfect condition and it looks as though this journey has just happened.

However, with the high resolution capability of the LROC, scientists were able to work out when they believed the journey took place. It is thought the boulder actually rolled between 50 and 100 million years ago.

Lunar scientist James Ashley said on the LROC website: "Studies suggest that regolith development from micrometeorite impacts will erase tracks like these over time intervals of tens of millions of years. Eventually it's track will be erased completely."

It is not known what might have caused the rock to roll, but Ashley has suggested the boulder moved because of impact and that this might have been caused by a direct hit from a meteoroid.

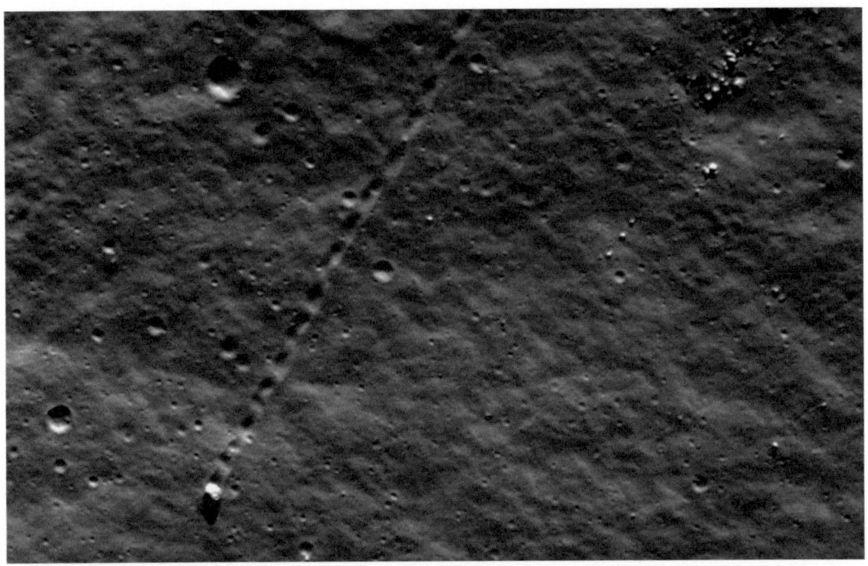

Fig. 6.8 The rolling stone is about the size of a truck. (Credit: NASA)

7. Waiter, There's Metal in My Moon Water!

Bring a filter if you plan on drinking water from the Moon. Water-ice recently discovered in dust at the bottom of a crater near the Moon's South Pole is accompanied by metallic elements such as mercury, magnesium, calcium, and even a bit of silver. Now you can add sodium to the mix, according to Dr. Rosemary Killen of NASA's Goddard Space Flight Center in Greenbelt, MD (Fig. 7.1).

Recent discoveries of significant deposits of water on the Moon were surprising because our Moon has had a tough life. Intense asteroid bombardments in its youth, coupled with its weak gravity and the Sun's powerful radiation, have left the Moon with almost no atmosphere. This rendered the lunar surface barren and dry, compared to Earth.

However, due to the Moon's orientation to the Sun, scientists theorized that deep craters at the lunar poles would be in permanent shadow and thus extremely cold, and able to trap volatile material like water as ice if such material were somehow transported there, perhaps by comet impacts or chemical reactions with hydrogen, a major component of the solar wind.

The impact of NASA's Lunar CRater Observation and Sensing Satellite (LCROSS) spacecraft into the permanently shadowed region of the Cabeus Crater confirmed that a surprisingly large amount of water-ice exists in this region, along with small amounts of many other elements, including metallic ones.

LCROSS was launched June 18, 2009, as a companion mission to NASA's Lunar Reconnaissance Orbiter, or LRO, from NASA's Kennedy Space Center in Florida. After separating from LRO, the LCROSS spacecraft held onto the spent Centaur upper stage rocket of the launch vehicle, executed a lunar swing-by, and entered into a series of long looping orbits around Earth.

After traveling for approximately 113 days and nearly 5.6 million miles (9 million km), the Centaur and LCROSS separated on

© Springer International Publishing Switzerland 2016

V.S. Foster, *Modern Mysteries of the Moon*, Astronomers' Universe,

DOI 10.1007/978-3-319-22120-5_7

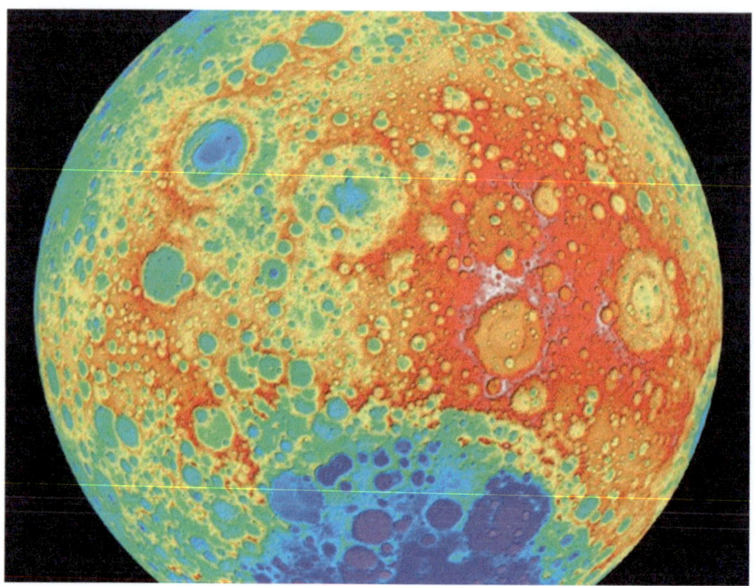

FIG. 7.1 As seen is this false-color image, the Moon is comprised of a wide range of resources, including water-ice, and various metals and mineral resources. All of these would be invaluable to whichever nation travels to the Moon next. (Credit: NASA)

final approach to the Moon. Moving faster than most rifle bullets, the Centaur impacted the lunar surface with LCROSS and LRO watching, using their onboard instruments. Approximately 4 min of data were collected by LCROSS before the spacecraft itself impacted the lunar surface.

Killen and her team observed the LCROSS impacts with the National Solar Observatory's McMath-Pierce solar telescope at the Kitt Peak National Observatory, Tucson, AZ. They were the only team able to see the results of the impacts from the ground.

The impacts vaporized volatile material from the bottom of Cabeus Crater, including water and sodium. After the vapor plume rose about 800 m (around 2600 ft.)—high enough to clear the shadow from the crater rim—sunlight stimulated the sodium atoms, causing them to emit their signature yellow–orange glow. A high-resolution Echelle spectrometer attached to the telescope detected this unique glow. The instrument separates light into its component colors to identify materials by the characteristic colors they emit when energized by radiation or other events in space.

The spectrometer views the sky through a narrow slit to separate the colors, so the team had to make assumptions about the shape and temperature of the plume to estimate the total amount of sodium liberated by the impacts. Using a computer model of the impact and other data on the impacts from instruments on LCROSS and LRO to guide their assumptions, the team calculated that about 1–2 kg (about 2.2–4.4 lbs.) of sodium were released. "This is 1–2 % of the amount of water released by the impacts," said Killen. "Our oceans have a comparable sodium to water ratio—about 1 %." (The amount of sodium derived from the observations depends on the assumed temperature of the vapor.)

This much sodium raises the question: where did it all come from? Sodium atoms from comet impacts could bounce across the lunar surface until they landed in the permanently shadowed regions, where they would get "cold trapped"—frozen in place. The solar wind carries small amounts of sodium, which could become embedded in the lunar surface, and it might also liberate sodium from lunar rocks, which are about 0.4 % sodium. Sodium is also liberated from lunar rocks by meteoroid impacts. (The LCROSS impacts didn't have enough energy to vaporize rock, so it's unlikely the sodium vapor plume simply came from rocks at the impact site, according to Killen.)

"Two percent sodium to water is consistent with the amount of sodium in comets, so perhaps the bulk of the sodium and water came from comet impacts," said Killen. She makes it clear that this is just speculation at this point, and that it's possible they came from a different source or even a variety of sources, including cold-trapped lunar volatiles and solar wind-induced chemistry. Better evidence for a cometary origin would come from an analysis of the hydrogen isotopes in lunar water, according to Killen.

Isotopes are versions of an element with different weights, or masses. For example, a deuterium atom is a heavier version of a common hydrogen atom because it has an extra particle—a neutron—in its nucleus at the center. Deuterium can be substituted for the regular form of hydrogen in a water molecule, but it is much less common than hydrogen, and its concentration varies in objects across the Solar System. If the deuterium to hydrogen ratio in lunar water is similar to the ratio in comets, it would suggest the water came from comet impacts. Since comets as "dirty snowballs" carry many other materials, it would imply that much of the sodium and other volatiles came from comets as well.

The team plans to shed light on the origin of lunar water and other volatiles using data from the Lunar Atmosphere and Dust Environment Explorer (LADEE) mission, which was launched in May 2013. The mission is orbiting the Moon and observing its tenuous atmosphere (technically called an exosphere, because it is so thin that atoms rarely collide with each other above the surface).

The research was funded by NASA's Dynamic Response of the Environment At the Moon (DREAM) project. "This discovery highlights a particular value of the DREAM program—we can rapidly support missions like the LCROSS impact with additional observations and analysis," said Dr. William Farrell of NASA Goddard, lead of the DREAM institute.

The McMath-Pierce telescope is operated by the National Solar Observatory, which is funded by the National Science Foundation and managed by the Association of Universities for Research in Astronomy. Killen's paper on this research was published in *Geophysical Research Letters* November 2014.

New Mineral Proves Old Idea About Space Weathering

Discovered in a lunar meteorite, a new mineral named hapkeite honors the scientist, Bruce Hapke (Emeritis Professor at University of Pittsburgh), who nearly 30 years ago predicted the importance of vaporization as one of the processes in space weathering. The new iron silicide mineral (Fe_2Si) was announced by the research team of Mahesh Anand (formerly at the University of Tennessee, Knoxville, and now at the Natural History Museum, London), Larry Taylor (University of Tennessee, Knoxville), Mikhail Nazarov (Vernadsky Institute of Geochemistry and Analytical Chemistry, Moscow), Jinfu Shu, Ho-kwang Mao, and Russell Hemley (Carnegie Institution of Washington).

This mineral likely formed by impact vaporization of the lunar soil and subsequent condensation of the iron and silicon into tiny metal grains. The researchers conclude that Fe-Si phases are more common in the lunar soil than previously thought. It is nanophase-sized FeO. These Fe-Si phases, and other space weathering products that profoundly affect the optical properties of the lunar soil at visible and near infrared wavelengths, must be taken into account when interpreting remote sensing data of the Moon.

The countless craters and basins we see on the Moon attest to a long history of meteorite bombardment. It's a bombardment rate that reached a maximum about 3.85 billion years ago—based on the ages of highland breccia samples returned by the Apollo astronauts. Since then, a lesser but continuous barrage of micrometeorites and charged particles from the Sun and stars has been generating a powdery surface, also called regolith (or soil, though no biologic component is implied in the name). With nothing to stop or slow down the incoming space debris, due to a near-complete lack of any atmosphere on the Moon, even the tiniest grains (mostly 10s to 150 µm) hit the Moon's surface at full cosmic velocity, 20 or more km per second (about 70,000 to 150,000 km/h).

The pulverized powdery regolith is just what you'd expect from impact debris. It has rock and mineral fragments, impact glasses, and particles called agglutinates that are mineral and rock fragments stuck together by impact glass. And, the high-energy micrometeorite impacts cause flash-melting, vaporization, and re-deposition of some of the surface materials. The phenomenon responsible for the physical and chemical makeup of the Moon's surface is utterly unique to airless planetary bodies—space weathering.

Space weathering is generally defined as the processes (such as meteorite, micrometeorite, and cosmic dust bombardment, solar wind ion implantation, and sputtering, but also including deep vacuum and temperatures approaching absolute zero [−273 °C) that change the physical structure, optical properties, and chemical and mineralogical properties of the surface of an airless planetary body from their original conditions. It applies to the Moon, Mercury, and asteroids, as well as other small bodies such as Phobos and Deimos.

The concept blossomed in the early 1970s when cosmochemists studying rock and regolith samples returned by the Apollo astronauts compared what they were finding with what was already known of the Moon's surface through telescopic spectral studies. There were some surprise results in the laboratories!

Apollo regolith samples were darker and spectrally redder than freshly crushed lunar rocks of similar composition.

Bruce Hapke (Emeritis Professor at University of Pittsburgh) and colleagues worked on the question of why the lunar soil becomes darker and spectrally redder with time. In 1975, they proposed that metallic iron would be among the elements and compounds vaporized during space weathering and subsequently condensed in glassy coatings on the surfaces of surrounding soil grains. These iron droplets would be minute, only nanometers in size (billionths of a meter), but they would dramatically alter how light interacts with the surface materials and, hence, how it would be sensed remotely. At the time of the predictions, however, no one could find any trace of vapor condensates on the lunar soil particles.

Finally, technology caught up with human inspiration. In 1993, Lindsay Keller and David McKay (Johnson Space Center) using transmission electron microscopy (TEM) found nanophase-sized metallic iron beads in silica-rich glassy rims on individual mineral grains in Apollo lunar samples. In the same year, Carlé Pieters (Brown University) and colleagues determined that the greatest spectral changes due to space weathering were in the smallest grains of the lunar regolith and concluded that they were caused by changes on the surfaces of grains rather than within the grains.

The general consensus reached in the past few years is that no single space-weathering phenomenon entirely explains the darkening and spectral-reddening characteristics of mature lunar regolith. Rather, processes work in concert (grain fragmentation or damage due to impact, agglutinate formation, solar-wind sputtering or impact vaporization and condensation of nanophase-sized iron) to alter the optical properties of the lunar regolith, with vapor-phase deposition of nanophase-sized iron playing the most significant role. Metallic iron particles not associated with grain coatings have now been identified within a lunar meteorite. The host rock of the newly discovered mineral is the lunar meteorite Dhofar 280, collected in 2001 in the Dhofar region of Oman. It is classified as an anorthositic fragmental breccia.

One of the breccia clasts in the meteorite contains opaque minerals 2–30 μm in size that have a slight tarnished appearance.

Upon closer inspection with the electron microprobe, Mahesh Anand and Larry Taylor discovered that the grains are actually compounds of iron silicides. The largest such grain (~35 μm) turned out to be Fe_2Si, and its discovery in Dhofar 280 is the first documentation of its natural occurrence.

Similar chemical structures have been created synthetically in the lab, and other related minerals have been known to form on Earth under unique conditions when lightning strikes sandy soil, forming glassy fulgurites. Naming the new Fe_2Si mineral hapkeite seemed fitting to Larry Taylor and his colleagues, as they consider it to be a direct product of impact-induced vapor-phase deposition in the lunar regolith. Hapkeite is the third iron silicide identified in this lunar meteorite.

The other two minerals remain to be formerly named, FeSi and $FeSi_2$. Anand and colleagues consider SiO_2 in the vapor phase from energetic impact-induced melts to be an important source of SiO_2+ and SiO, which would combine in various proportions in the vapor with FeO to condense out as the observed Fe-Si metal grains. Although hapkeite has not yet been identified in Apollo regolith samples, the research team concluded that Fe-Si phases are probably more common in the lunar regolith and may be more closely related in origin to nanophase iron than previously thought. As the *Lunar Sourcebook* (p. 286) says, "The regolith is the source of virtually all our information about the Moon." Covering practically the entire surface to a depth of about 3–10 m, the regolith is the source of our Apollo samples and dominates what our remote sensing instruments analyze. Since Hapke's predictions, cosmochemical analyses of samples have helped to explain how the formation and accumulation of nanophase iron and Fe-Si phases have changed the physical, optical, chemical, and mineralogical properties of the lunar regolith.

Much of what we know today about the global properties of the Moon stems from early work by scientists such as Hapke, Tom McCord (Emeritis Professor at the University of Hawaii), and their colleagues on the effects of vapor-phase depositional processes and the effects on the chemical and optical properties of the lunar regolith. Building on this knowledge of space weathering, researchers are developing new mineral mapping techniques in visible and near infrared wavelengths. For example, Paul Lucey

(University of Hawaii) and colleagues have used Clementine multispectral data to determine the concentrations of FeO and TiO_2 on the lunar surface.

Experts in lunar sample analysis are collaborating with experts in remote sensing to address outstanding questions in lunar science. One such group, for example, the Lunar Soil Characterization Consortium, is working to better understand the effects of space weathering on the surfaces of airless planetary bodies. They are characterizing the mineralogy and chemistry of the finest-sized fractions of the lunar regolith in order to better understand remotely sensed reflectance spectra of the Moon.

The discovery of hapkeite in a lunar meteorite has helped improve our understanding of space weathering on the Moon and how space weathering plays a major role in affecting remote sensing studies of airless planetary bodies.

There's Gold in Them Thar Hills

New research reveals that the abundance of so-called highly siderophile, or metal-loving, elements such as gold and platinum found in the mantles of Earth, the Moon and Mars were delivered by massive impactors during the final phase of planet formation over 4.5 billion years ago. The predicted sizes of the projectiles, which hit within tens of millions of years of the giant impact that produced our Moon, are consistent with current planet formation models as well as physical evidence such as the size distributions of asteroids and ancient Martian impact scars.

They predict that the largest of the late impactors on Earth—at 1500–2000 miles in diameter—potentially modified Earth's obliquity by approximately 10°, while those for the Moon, at approximately 150–200 miles, may have delivered water to its mantle.

The team that conducted this study was comprised of Solar System dynamicists, such as Dr. William Bottke and Dr. David Nesvorny from the Southwest Research Institute, and geophysical-geochemical modelers, such as Prof. Richard J. Walker from the University of Maryland, Prof. James Day from the University of Maryland and Scripps Institution of Oceanography, and Prof. Linda

Elkins-Tanton, from Arizona State University. Together, they represent three teams within the NASA Lunar Science Institute (NLSI).

A fundamental problem in planetary science is to determine how Earth, the Moon, and other inner Solar System planets formed and evolved. This is a difficult question to answer given that billions of years of history have steadily erased evidence for these early events. Despite this, critical clues can still be found to help determine what happened, provided one knows where to look.

For instance, careful study of lunar samples brought back by the Apollo astronauts, combined with numerical modeling work, indicates that the Moon formed as a result of a collision between a Mars-sized body and the early Earth about 4.5 billion years ago. While the idea that the Earth-Moon system owes its existence to a single, random event was initially viewed as radical, it is now believed that such large impacts were commonplace during the end stages of planet formation. The giant impact is believed to have led to a final phase of core formation and global magma oceans on both Earth and the Moon.

For the giant impact hypothesis to be correct, one might expect samples from the Earth's and Moon's mantle, brought to the surface by volcanic activity, to back it up. In particular, scientists have examined the abundance in these rocks of so-called highly siderophile, or metal-loving, elements: Re, Os, Ir, Ru, Pt, Rh, Pd, Au. These elements should have followed the iron and other metals to the core in the aftermath of the Moon-forming event, leaving the rocky crusts and mantles of these bodies void of these elements. Accordingly, their near-absence from mantle rocks should provide a key test of the giant impact model.

However, as described by team member Walker, "The big problem for the modelers is that these metals are not missing at all, but instead are modestly plentiful." Team member Day adds, "This is a good thing for anyone who likes their gold wedding rings or the cleaner air provided by the palladium in their car's catalytic convertors."

A proposed solution to this conundrum is that highly siderophile elements were indeed stripped from the mantle by the effects of the giant impact, but were then partially replenished by later impacts from the original building blocks of the planets, called

Fig. 7.2 NASA is collaborating with the European and Canadian space agencies to develop technologies that could, one day, be used to manufacture oxygen and other vital resources out of the lunar regolith (soil). (Credit: NASA)

planetesimals. This is not a surprise—planet formation models predict such late impacts—but their nature, numbers, and most especially size of the accreting bodies are unknown. Presumably, they could have represented the accretion of many small bodies or a few large events. To match observations, the late-arriving planetesimals needed to deliver 0.5 % of Earth's mass to Earth's mantle, equivalent to one-third of the mass of the Moon, and about 1200 times less mass to the Moon's mantle (Fig. 7.2).

Using numerical models, the team showed that they could reproduce these amounts if the late accretion population was dominated by massive projectiles. Their results indicate the largest Earth impactor was 1500–2000 miles in diameter, roughly the size of Pluto, while those hitting the Moon were only 150–200 miles across. Lead author Bottke says, "These impactors are thought to be large enough to produce the observed enrichments in highly siderophile elements, but not so large that their fragmented cores joined with the planet's core. They probably represent the largest objects to hit those worlds since the giant impact that formed our Moon."

Intriguingly, the predicted distribution of projectile sizes, where most of the mass of the population is found among the largest objects, is consistent with other evidence.

New models describing how planetesimals form and evolve suggest the biggest ones efficiently gobble up the smaller ones and run away in terms of size, leaving behind a population of enormous objects largely resistant to collisional erosion.

The last surviving planetesimal populations in the inner Solar System are the asteroids. In the inner Asteroid Belt, the asteroids Ceres, Pallas, and Vesta, at 600 miles, 300 miles, and 300 miles across, respectively, dwarf the next largest asteroids at 150 miles across. No asteroids with "in-between" sizes are observed in this region.

The sizes of the oldest and largest craters on Mars, many of which are thousands of miles across, are consistent with it being bombarded by an inner Asteroid Belt-like population dominated by large bodies early in its history.

These results make it possible to make some interesting predictions about the evolution of Earth, Mars and the Moon. For example, the largest projectiles that struck Earth were capable of modifying its spin axis, on average, by approximately 10°.

Moon May Harbor Alien Minerals

Minerals found in craters on the Moon may be remnants of asteroids that slammed into it and not, as long believed, the satellite's innards exposed by such impacts, one study said.

The findings, published in the journal *Nature Geoscience*, cast doubt on the little we knew of what the Moon is actually composed of. It had long been thought that meteoroids vaporize on impact with large celestial bodies. Unusual minerals such as spinel and olivine found in many lunar craters, but rarely on the Moon's surface, were therefore attributed to the excavation of subsurface lunar layers by asteroid hits.

FIG. 7.3 Moon craters may hold minerals from ancient asteroids, computer models suggest. (Credit: NASA)

Olivine and spinel are common components of asteroids and meteorites, and have been found on the floors and around the central peaks of lunar craters such as Copernicus, Theophilus, and Tycho that are around 100 km (63 miles) in diameter.

A team from China and the United States simulated the formation of Moon craters and found that at impact velocities under 12 km per second a projectile may survive the impact, though fragmented and deformed (Fig. 7.3).

"We conclude that some unusual minerals observed in the central peaks of many lunar impact craters could be exogenic (external) in origin and may not be indigenous to the Moon," they wrote.

Co-author Jay Melosh from Purdue University in Indiana, said the finding answers the conundrum exposed by earlier studies that said craters the size of Copernicus were not big enough to have dredged up the contents of the Moon's deep, interior mantle.

"It also warns planetary scientists not to use the composition of the central peaks of craters as a guide to the interior of the Moon, whose dominant mineral might not be olivine," he said. On Earth, spinel and olivine create rare gemstones such as peridot.

In an article commenting on the study, Erik Asphaug of the School of Earth and Space Exploration at Arizona State University, said the theory meant that material excavated from Earth by large impacts during the planet's early days may still be found on the Moon.

The scattered material was known to have hit the Moon at velocities as slow as 2 km/s and should have survived, if the study's assumptions are correct. This suggested yet another explanation for the existence of spinels on the Moon, said Asphaug—they came from Earth.

"Even more provocative is the suggestion that we might someday find Earth's protobiological materials, no longer available on our geologically active and repeatedly recycled planet, in dry storage up in the lunar 'attic.'

"Certainly, the potential of finding early Earth material is emerging as one of the primary motivations for a return to the Moon by human astronauts in our ongoing search for the origin of life."

Unlike Earth's crust, which is repeatedly recycled through the process of plate tectonics, the Moon's hard crust dates back billions of years, offering clues to the formation of the Solar System, including Earth.

Scientists Find Silver, Too

Scientists led by Brown University are offering the first detailed explanation of the crater formed when a NASA rocket slammed into the Moon and information about the composition of the lunar soil at the poles that never has been sampled. The findings are published in a set of papers in *Science* stemming from the successful NASA mission, LCROSS.

Mission control at NASA Ames sent the emptied upper stage of a rocket crashing into the Cabeus crater near the Moon's South Polee in October 2009. A second spacecraft followed to analyze the ejected debris for signs of water and other constituents of the super-chilled lunar landscape.

In one of the papers, Brown planetary geologist Peter Schultz and graduate student Brendan Hermalyn, along with NASA scientists, write that the cloud kicked up by the rocket's impact showed the Moon's soil and subsurface is more complex than believed. Not only did the lunar regolith—the soil—contain water, it also harbored other compounds, such as hydroxyl, carbon monoxide, carbon dioxide, ammonia, free sodium, and, in a surprise, silver.

Combined, the assortment of volatiles—the chemical elements weakly attached to regolith grains—gives scientists clues where they came from and how they got to the polar craters, many of which haven't seen sunlight for billions of years and are among the coldest spots in the Solar System.

Schultz, lead author on a *Science* paper, thinks many of the volatiles originated with the billions of years-long fusillade of comets, asteroids and meteoroids that have pummeled the Moon. He thinks an assortment of elements and compounds, deposited in the regolith all over the Moon, could have been quickly liberated by later small impacts or could have been heated by the Sun, supplying them with energy to escape and move around until they reached the poles, where they became trapped beneath the shadows of the frigid craters.

"This place [Cabeus Crater] looks like it's a treasure chest of elements, of compounds that have been released all over the Moon," Schultz said, "and they've been put in this bucket in the permanent shadows."

Schultz believes the variety of volatiles found in Cabeus Crater's soil implies a kind of tug of war between what is being accumulated and what is being lost to the tenuous lunar atmosphere.

"There's a balance between delivery and removal," explained Schultz, who has been on the Brown faculty since 1984 and has been studying the Moon since the 1960s. "This suggests the delivery is winning. We're collecting material, not simply getting rid of it."

Astronauts sent as part of NASA's Apollo missions found trace amounts of silver, along with gold, on the near side (Earth-facing side) of the Moon. The discovery of silver at Cabeus Crater suggests that silver atoms throughout the Moon migrated to the poles. Nevertheless, the concentration detected from Cabeus "doesn't mean we can go mining for it," Schultz said.

The crater formed by the rocket's impact within Cabeus produced a hole 70–100 ft in diameter and tossed up 6-ft-deep lunar material. The plume of debris kicked up by the impact reached more than a half-mile above the floor of Cabeus, high enough to rise into sunlight, where its properties could be measured for almost 4 min by a variety of spectroscopic instruments. The amount of ejecta measured was almost 2 t, the scientists report. The scientists also noted there was a slight delay, lasting roughly one-third of a second, in the flash generated after the collision. This indicated to them that the surface struck may be different than the loose, almost crunchy surface trod by the Apollo astronauts.

"If it had been simply lunar dust, then it would have heated up immediately and brightened immediately," Schultz said. "But this didn't happen."

The scientists also noticed a one-half-mile near-vertical column of ejecta still returning to the surface. Even better, the LCROSS spacecraft was able to observe the plume as it followed on the heels of the crashing rocket. Schultz and Hermalyn had observed such a plume when conducting crater-impact experiments using hollow spheres (that mimicked the rocket that crashed into Cabeus) at the NASA Ames Vertical Gun Range in California before the LCROSS impact.

"This was not your ordinary impact," Hermalyn said. "So in order to understand what we were going to see (with LCROSS) and maybe what effects that would have on the results, we had to do all these different experiments."

Even though the mission has been judged a success, Schultz said it posed at least as many questions as it answered (Fig. 7.4).

"There's this archive of billions of years [in the Moon's permanently shadowed craters]," Schultz said. "There could be clues there to our Earth's history, our Solar System, our galaxy. And it's all just sitting there, this hidden history, just begging us to go back."

Contributing authors on the paper include Anthony Colaprete, Kimberly Ennico, Mark Shirley, and William Marshall, all from NASA Ames Research Center in California. NASA funded the research.

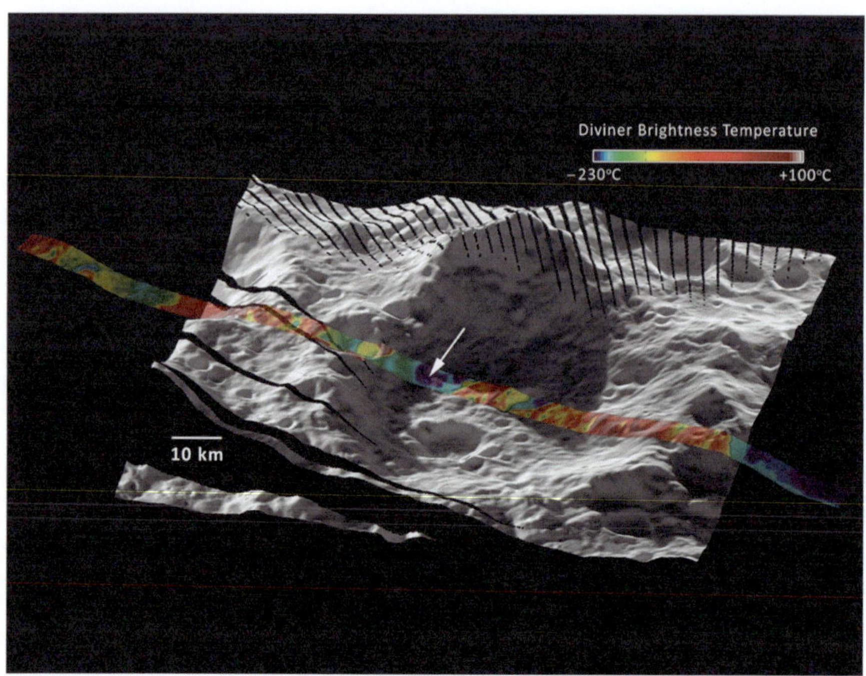

Fig. 7.4 Diviner brightness temperature swath acquired about 90 s after the LCROSS impact, the location of which is indicated by the *white arrow*. Based on the Diviner measurements, the impact site was heated to more than 700 °C (1300 °F). (Credit: NASA)

8. Strange Phenomena

Lunar Transient Phenomenon

A lunar transient phenomenon (LTP) is a short-lived light, glow, mist, obscuration, or some other change in appearance on the surface of the Moon. Claims of LTPs go back at least 1000 years, with some having been observed independently by multiple witnesses or reputable astronomers. Nevertheless, the majority of lunar transient phenomenon reports are not reproducible, nor do they possess adequate controls that could be used to distinguish them from among alternative hypotheses to explain their origins.

Thus, the professional community rarely studies these observations, and few reports concerning these phenomena are ever published in peer-reviewed scientific journals. Most lunar scientists will acknowledge that transient events such as outgassing and impact cratering do occur over geologic time. The controversy lies in the reporting of such events by individuals without corroborating evidence.

The term—lunar transient phenomenon—was created by Sir Patrick Moore during his co-authoring of NASA Technical Report R-277 *Chronological Catalog of Reported Lunar Events*, published in 1968 (Fig. 8.1).

Reports of LTPs range from foggy patches to color changes on the lunar landscape. Winifred S. Cameron, the world's top LTP expert, classifies these as (1) gaseous, involving mists and other forms of obscuration, (2) reddish colorations, (3) green, blue, or violet colorations, (4) brightenings, and (5) darkenings. Two extensive catalogs of transient lunar phenomena exist, with the most recent tallying 2254 events going back to the sixth century. At least one-third come from the vicinity of the Aristarchus plateau.

© Springer International Publishing Switzerland 2016

V.S. Foster, *Modern Mysteries of the Moon*, Astronomers' Universe,

DOI 10.1007/978-3-319-22120-5_8

FIG. 8.1 Past research and observations by amateurs, and by professional planetary scientists and astronomers, have proven that changes and effects have taken place on the Moon very recently, and are sometimes visible from Earth. These can take the form of short duration brightness changes, obscurations, colored glows and flashes of light. This photo of the Moon shows where many LTP have been observed. (Credit: NASA)

Among the most famous historical accounts of LTPs are the following:

- Five monks from Canterbury, England, reported an upheaval on the Moon shortly after sunset on June 18, 1178. "There was a bright new Moon, and as usual in that phase its horns were tilted toward the east; and suddenly the upper horn split in two. From the midpoint of this division a flaming torch sprang up, spewing out, over a considerable distance, fire, hot coals, and sparks. Meanwhile the body of the Moon which was below writhed, as it were, in anxiety, and, to put it in the words of those who reported it to me and saw it with their own eyes, the Moon throbbed like a wounded snake. Afterwards it resumed its proper state. This phenomenon was repeated a dozen times or more, the flame assuming various twisting shapes at random and then returning to normal. Then after these transformations the Moon from horn to horn, that is, along its whole length, took on a blackish appearance."

- In 1976, Jack Hartung proposed that the foregoing described the formation of the Giordano Bruno Crater. However, this appears highly unlikely, or even whether it was a true lunar transient phenomenon at all. An impact large enough to leave a 22-km-wide crater would have ejected millions of tons of lunar debris resulting in an unprecedented, week-long meteor storm on Earth. No account of such a storm have been found in any known historical records. As a result, it is suspected that the group of monks (the event's only known witnesses) saw the atmospheric explosion of an oncoming meteor in chance alignment with the distant Moon from their specific vantage point.

- British astronomer Sir William Herschel noticed three glowing red spots on the night side of the Moon on the night of April 19, 1787. He informed King George III and other astronomers of his observations. Herschel pointed to erupting volcanoes as the cause and reported the luminosity of the brightest of the three as greater than the brightness of a comet he had observed on April 10. His observations were made while an aurora borealis (northern lights) rippled above his site in Padua, Italy. It was very rare for aurora activity to occur that far south from the Arctic Circle, and Padua's display and Herschel's observations had happened just a few days before the number of sunspots had peaked.

- The experienced lunar observer and mapmaker J. F. Julius Schmidt claimed in 1866 that the Linné Crater had changed its appearance. Based on drawings made earlier by J. H. Schröter, as well as personal observations and drawings made between 1841 and 1843, he stated that the crater "at the time of oblique illumination *cannot at all be seen* (his emphasis), whereas at high illumination, it was visible as a bright spot." Based on repeat observations, he further stated that "Linné can never be seen under any illumination as a crater of the normal type" and that "a local change has taken place." Today, Linné is visible as a normal young impact crater with a diameter of about 1.5 miles (2.4 km).

- On November 2, 1958, the Russian astronomer Nikolai A. Kozyrev, using a 48-in. (122-cm) reflector telescope equipped with a spectrometer, observed an apparent half-hour "eruption" that took place on the central peak of Alphonsus Crater. Spectra showed evidence for bright gaseous emission bands for the molecules C_2 and C_3. While exposing his second spectrogram, he noticed "a marked increase in the brightness of the central region and an unusual white color." Then, "all of a sudden the brightness started to decrease," and the resulting spectrum was normal.

- On October 29, 1963, James Clarke Greenacre and Edward M. Barr, two observers at the Lowell Observatory, Flagstaff, AZ, manually recorded very bright red, orange, and pink coloring on the southwestern side of Cobra Head; a hill southeast of the lunar valley Vallis Schröteri; and the southwestern interior rim of the Aristarchus Crater. This event sparked a major change in attitude towards LTP reports. According to Willy Ley: "The first reaction in professional circles was, naturally, surprise, and hard on the heels of this followed an apologetic attitude, the apologies being directed at a long-dead great astronomer, Sir William Herschel. The credibility of their findings stemmed from Greenacre's exemplary reputation as an impeccable cartographer, rather than from any photographic evidence.

- A few days after the Greenacre event, Zdenek Kopal and Thomas Rackham, at the Observatoire du Pic-du-Midi in the French Pyrenees, made the first photographs of a "wide area of lunar luminescence." Kopal's article in *Scientific American* transformed it into one of the most widely publicized LTP events. Kopal, like others, had argued that energetic solar particles could be the cause of such a phenomenon.

- One evening near the end of 1992, Audouin Dollfus of the Observatoire de Paris reported anomalous features on the floor of Langrenus Crater using a 1-m (3.2-ft) telescope. Unusually high albedo and polarization features were recorded that did not change in appearance over the 6 min of data collection. Three days later observations showed a similar, but smaller, anomaly

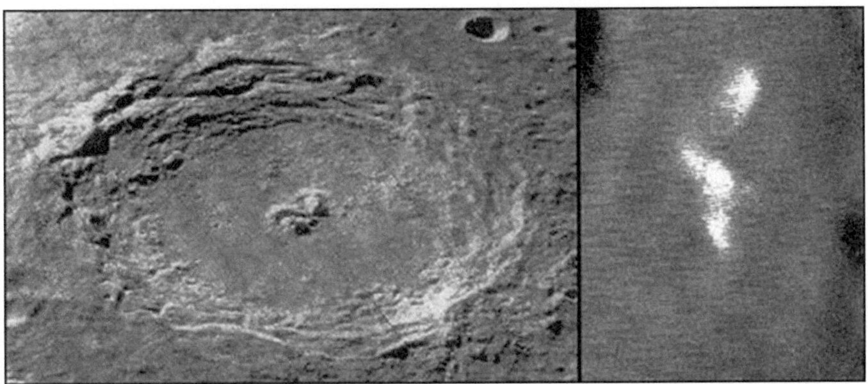

FIG. 8.2 In 1992, the veteran lunar observer Audouin Dollfus saw a series of glows on the floor of the crater Langrenus using the 1-m (39-in.) Meudon reflecting telescope of the Observatoire de Paris. Dollfus believes that the glows were due to escaping gas that lifted dust above the lunar surface into the sunlight. He pointed out that Langrenus, when observed in detail, had an extensive series of fractures on its crater floor, and the gas could have been escaping from these. In 1994, the Clementine lunar orbiter observed Aristarchus before and after a LTP was seen from Earth. Spectral data from the probe suggested that parts of the crater had changed color slightly. Despite the various evidence that has accrued, the nature and even the existence of LTPs as real physical phenomena remain open issues. (Credit: NASA)

in the same vicinity. While the viewing conditions for this region were close to specular, it was argued that the amplitude of the observations were not consistent with a specular reflection of sunlight. The favored hypothesis was that this was the consequence of light scattering from clouds of airborne particles resulting from a release of gas. The fractured floor of this crater was cited as a possible source of the gas (Fig. 8.2).

The cause of lunar transient phenomena generally fall into four classes: outgassing, impact events, electrostatic phenomena, and unfavorable observing conditions.

Some LTPs may be caused by gas escaping from underground cavities. These events are purported to display a distinctive reddish hue, while others have appeared as white clouds or an indistinct haze. The majority of LTPs appear to be associated with floor-fractured craters, the edges of lunar maria, or in other locations linked by geologists with volcanic activity. However, these are some of the most common observing targets by amateur astronomers, and this correlation could have an observational bias.

In support of the outgassing hypothesis, data from the Lunar Prospector alpha particle spectrometer indicated the recent outgassing of radon to the surface. In particular, results show that radon gas was emanating from the vicinity of the craters Aristarchus and Kepler during the time of this 2-year mission. These observations could be explained by the slow and visually imperceptible diffusion of gas to the surface, or by discrete explosive events. In support of explosive outgassing, it has been suggested that a roughly 3-km (1.9 miles)-diameter region of the lunar surface was "recently" modified by a gas release event. The age of this feature is believed to be about 1 million years old, suggesting that such large phenomena occur only infrequently.

Impact events occur continually on the lunar surface. The most common events are those associated with micrometeorites, as might be encountered during meteor showers. Impact flashes from such events have been detected from multiple and simultaneous Earth-based observations. Furthermore, impact clouds were detected following the crash of ESA's SMART-1 spacecraft, India's Moon Impact Probe, and NASA's LCROSS. Impact events leave a visible scar on the surface, and these could be detected by analyzing before and after photos of sufficiently high resolution. No impact craters formed between the Apollo-era, Clementine (global resolution 100 m, selected areas 7–20 m) and SMART-1 (resolution 50 m) missions have been identified.

Electrostatic charging or discharging might be able to account for some LTP. One possibility is that electrodynamic effects related to the fracturing of near-surface materials could charge any gases that might be present, such as implanted solar wind or radiogenic daughter products. If this were to occur at the surface, the subsequent discharge from this gas might be able to give rise to phenomena visible from Earth. Alternatively, it has been proposed that the triboelectric charging of particles within a gas-borne dust cloud could give rise to electrostatic discharges. Finally, electrostatic levitation of dust near the terminator could potentially give rise to some type of phenomenon.

It is possible that many transient phenomena might not be associated with the Moon itself but could be a result of unfavorable observing conditions or phenomena associated with Earth. For instance, some reported transient phenomena are for objects near the top resolution of a particular telescope. Earth's atmosphere can give rise to significant temporal distortions that could be confused with actual lunar phenomena (an effect known as astronomical seeing). Other non-lunar explanations include the viewing of Earth-orbiting satellites and meteors or observational error.

The most significant problem that faces reports of LTPs is that the vast majority of these were made either by a single observer or at a single location on Earth (or both). The multitude of reports for transient phenomena occurring at the same *place* on the Moon could be used as evidence supporting their existence. Events must be regarded with caution in the absence of eyewitness reports from multiple observers at multiple locations on Earth for the *same* event. As discussed above, an equally plausible hypothesis for the majority of these events is that they are caused by the terrestrial atmosphere. If an event were to be observed at two different places on Earth at the same time, this could be used as evidence against an atmospheric origin.

One attempt to overcome the above problems with transient phenomena reports was made during the Clementine mission by a network of amateur astronomers. Several events were reported, of which four of these were photographed both beforehand and afterward by the spacecraft. However, careful analysis of these images shows no discernible differences at these sites. This does not necessarily imply that these reports were a result of observational error, as it is possible that outgassing events on the lunar surface might not leave a visible marker, but neither is it encouraging for the hypothesis that these were authentic lunar phenomena.

Bill Cooke of NASA's Marshall Space Flight Center has a set of robotic telescopes that he is using to monitor the night side of the Moon for impact flashes. To date over 100 have been recorded on low-light sensitive video cameras. Arlin Crotts of Columbia University has been operating a robotic telescope in Cerro Tololo, Chile, and other worldwide locations to monitor the Moon's surface in white light, every few seconds to look for changes. He suggests that many LTPs are caused by releases of radon gas during moonquakes that erupt from the surface through cracks. This raises dust temporarily and increases reflectivity in the short term.

His team makes ~300 h of observations per month but have yet to analyze their results.

Efforts are currently being made by the Association of Lunar and Planetary Observers and the British Astronomical Association to re-observe sites where lunar transient phenomena were reported in the past. By documenting the appearance of these features under the same illumination and libration conditions, it is possible to judge whether some reports were simply due to a misinterpretation of what the observer regarded as an abnormality. Furthermore, with digital images, it is possible to simulate atmospheric spectral dispersion, astronomical seeing blur, and light scattering by our atmosphere to determine if these phenomena could explain some of the original LTP reports.

Following are guidelines offered by the British Astronomical Association and American Lunar and Planetary Observers for regularly observing LTPs.

Observing Program 1

Make detailed sketches, or high resolution images, of certain lunar features under the same illumination and/or libration to what they were like in past LTP observations. These repeat illumination/libration observations of a feature can tell us whether what was reported as a LTP was normal or not.

Observing Program 2

Look for impact flashes on the Moon's night side. Using low-light-level CCTV cameras, such as those used typically for occultation timings, in order to record <0.1 s duration impact flashes. Using integrating CCD cameras, take time lapse images of Earthshine, to look for evidence of debris from impacts making it into sunlight and becoming visible.

Observing Program 3

Monitor Ina-type features and their surroundings frequently. Pete Schultz of Brown University has proposed that Ina structures on the lunar surface are geologically recent and so might still occasionally outgas. Observe such areas closely for evidence of colored glows, brightenings, and obscurations, which might indicate outgassing.

Observing Program 4

Watch the terminator and crater shadow interiors for glows. Look for evidence of colored glows from radon gas being ionized in sunlight, as seen against a dark background. Also, be on the lookout for evidence of electrostatically levitated dust particles on the terminator, again utilizing the dark background to help detect these. Polaroid filters may help here.

Observing Program 5

Use time-lapse video to monitor large areas of the lunar surface. This can replicate Arlin Crotts' work except that this can be done through narrowband filters, specific to emission lines that one might expect from outgassing, e.g., radon at 705 nm, 745 nm, 810 nm, and 860 nm; argon at 764 nm, 811 nm, and 840 nm; sodium (which often contributes to the lunar atmosphere after an impact) at 589 nm; and hydrogen at 656 nm.

Observing Program 6

Use an electronic moon-blink device to monitor the lunar surface visually. This is similar to the previous strategy, but the human observer can be brought into the loop for interactive detection of LTPs. Different narrowband filters can be switched into the optical path for, say, 1 s before moving onto the next.

With thousands of craters to choose from to observe for LTPs, you are probably wondering, "Where should I start?" Well, it has been found that a select number of lunar formations have a history of repeat events. This being the case, it narrows the search down to just 27 locations on the Moon where you can begin your search. These formations are as follows:

Lunar transient phenomena sites

Formations	Location	#LTP	BR	D	G	R	BL
Agrippa	4 N/11E	34	15	9	27	1	11
Alphonsus	13S/3W	46	19	9	17	27	2
Archimedes	30 N/4 W	5	2	1	2	0	0
Aristarchus	24 N/48 W	448	256	37	131	112	43
Atlas	47 N/44E	17	4	5	2	1	1
Censorinus	0/32E	11	10	0	2	2	1
Cobra Head	24 N/48 W	13	4	5	8	11	3
Copernicus	10 N/20 W	22	13	2	5	6	8
Mare Crisium	18 N/58E	27	14	7	10	5	3
Eratosthenes	15 N/11 W	16	12	2	5	2	2
Gassendi	18S/40W	33	6	4	9	24	0
Grimaldi	6S/68W	18	9	2	5	2	8
Herodotus	23 N/50 W	34	16	7	11	16	5
Kepler	8 N/38 W	17	14	1	0	3	3
Linnie	28 N/12E	19	7	8	6	1	0
Manilius	15 N/9E	14	12	1	4	4	3
Menelaus	16 N/16E	13	10	3	4	4	4
Picard	15 N/55E	15	12	2	3	2	2
Mons Pico	46 N/9 W	8	3	2	4	2	1
Mons Piton	41 N/1 W	10	3	1	8	8	0
Plato	51 N/9 W	114	53	13	51	31	13
Posidonius	32 N/30E	11	2	0	7	3	1
Proclus	16 N/47E	72	31	33	26	10	15
Schroter's Valley	26 N/52 W	25	4	0	20	5	1
Schickard	445S/26E	8	3	2	6	0	1
Theophilus	12S/26E	11	6	0	4	9	2
Tycho	43S/11W	16	12	0	6	1	2

BR Bright

D Dark

G Gaseous

R Red

Bl Blue

The information presented in this table represents data taken from the *Lunar Transient Phenomena Catalog*, NSSDC/WDC-A-R&S 78–03. The catalog was published by National Space Science Data Center (NSSDC)/World Data Center A for Rockets and Satellites (WDC-A-R&S, National Aeronautics and Space Administration, Goddard Space Flight Center, Greenbelt, Maryland 20771). It covers time periods from A.D. 500 to May 28, 1977, and was authored by Winifred S. Cameron.

Donut Craters

The entire surface of the Moon is covered by craters, ranging from micron-sized pits to giant multi-ringed impact basins. One of the most unusual looking and least understood is the concentric crater found in only a few locations on the Moon. Looking like a donut inside of the bowl of a crater, it cannot be an impact that just happened to be centered on a preexisting crater (Fig. 8.3).

Usually, the size and morphology of a crater depend on the size and velocity of an impacting body. Crater size and form can be used to group lunar craters into three basic categories: simple craters, complex craters, and multi-ring basins. Simple craters are circular, bowl-shaped, and usually less than 10–15 km in diameter. Complex craters are usually larger than 10–20 km in diameter and have a well-defined central peak. The central peak shoots up

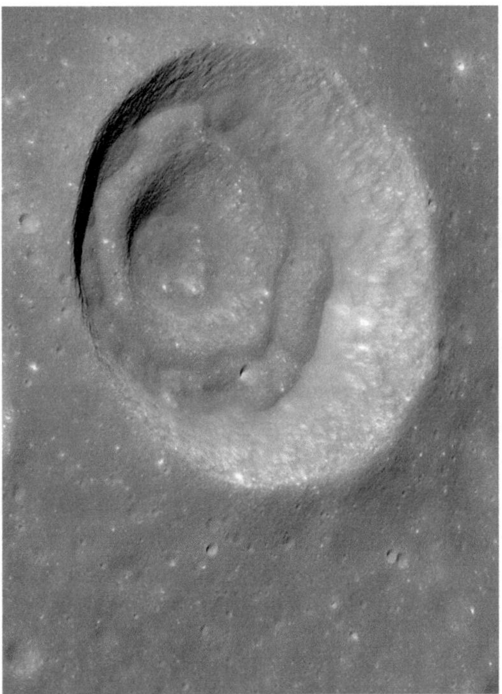

FIG. 8.3 LRO Wide Angle Camera (WAC) mosaic of the 11.5-km concentric crater (center), in context with the northern and northwestern Apollo basin. (Credit: NASA)

from great depths beneath the crater as the ground rebounds after the shock and pressure of the bolide impact. Complex craters also often have a rim with one or more terraces. Multi-ringed basins are remnants of the largest impacts on the lunar surface and usually have more than one rim, such as the 1100-km-diameter Imbrium Basin.

Some craters possess characteristics outside of these three basic categories: elliptical, polygonal, and concentric. Elliptical, or oval-shaped, craters form when the impactor hits the surface at an oblique angle, usually around 10° or less. Polygonal craters can form for a number of reasons, and the usual morphology results from pre-existing structural weaknesses in the material at the impact site.

Concentric is the third category in which the crater exhibits one or more donut-shaped scarps, called an inner torus. Normally, crater morphologies are attributed to impact processes. However, this does not easily explain the inner torus of concentric craters. Consequently, four models have been proposed for the formation of the inner torus: successive impacts, extrusive doming, mass wasting and impact into layered targets.

Measurements and observations were made using topographic profiles along with optical, near infrared, and radar images obtained from Clementine, Selene, and the LRO. Clementine-derived FeO and optical maturity maps also were used to examine these craters.

There are 57 known concentric craters. Their global distribution is similar to floor-fractured craters, usually located along the margins of mare regions or near isolated mare ponds. They are also typically isolated, rarely occurring within 100 km of each other. Concentric craters range from 3 to 20 km in diameter and average ~9 km. The diameter of the inner torus ranges from 1.5 to 9.7 km from the crest, with an average of ~4.3 km. In contrast, the diameter of floor-fractured craters ranges from 15 to 200 km and average ~40 km.

Three types of concentric craters have been identified. The first consists of a continuous smooth inner torus. The joint between the crater wall and the torus is a V-shaped valley. Another has a ring of roundish bumps in the moat between the inner and outer rims, sometimes referred to as "ball-bearings." The third

FIG. 8.4 Is Marth a concentric crater? It has a flat floor, unlike other concentric craters. The only significant elevation seems to be a tiny hill on the south side of the inner feature; most of the other inner depression seems to have only a very slight rim. The curved showdown of the inner feature looks like that of a typical small impact crater with a bright side and a dark side indicating a simple bowl shape. Marth looks more like two random impacts than like any known concentric crater. The image also shows that the small rille near Marth is slightly interrupted by the crater, but appears to continue on to the east. Just where the rille meets Marth it seems to affect the rim. This is strange!

is composed of a continuous torus except the joint between the crater wall and the torus is an inflection point (the point on a limb at which the concavity reverses). Some concentric craters contain a combination of all three types (Fig. 8.4).

To understand how concentric craters are different, they were compared to fresh, simple craters. As indicated by many scientists, these crater parameters, such as rim height, rim-flank width, and crater depth tend to be closely related to crater diameter. For example, the crater depth to diameter ratio is about 1:5 for fresh lunar craters. In comparing these different parameters to those of concentric craters it was found that concentric craters have shallower depths, shorter rim-flank widths, and smaller rim heights. This suggests concentric craters experienced degradation from impact and/or a combination of other mechanisms such as uplift of the crater floor.

Crater formation processes such as successive impacts and mass wasting do not explain the preferential distribution of concentric craters along mare-highland boundaries. An impact into layered targets require the interior of fresh craters near concentric craters to be modified as well; however, concentric craters are not found to cluster. Compositional maps do not show differences between the concentric crater and the local region, which is evidence against extrusion.

Planetary geologist Chuck Wood in his 1978 paper describes several variations of the morphology. Most common are typified by Hesiodius, with the usual characteristics as described above. Marth was said to appear like a cratered dome. (Marth looks more like two random impacts than any known concentric crater.) Some craters have elliptical inner and outer rings, as typified by the crater in Struve, while others have a round outer ring and elliptical toroid such as Crozier H, which is also off center from the outer ring. Craters such as Gruithuisen K have a raised moat area that resembles a third ring. A few craters are fractured, cracked, and highly worn, such as the concentric craters near Mons Jura, and have been described as having a "bread crust"-appearing interior. Garbart J is shown to have an inner rim that is low and inconspicuous, while Chamberlain was described as having a high concentric "collar," but on LRO images looks as if this could possibly be due to separate impacts (Fig. 8.5).

Any hypothesis for concentric crater formation must account for the following:

- the non-uniform distribution of the concentric craters
- the presence of normal non-concentric craters of younger age in close proximity to some of the concentric craters
- the presence of ejecta blanket remnants around several of the concentric craters
- shallow depth compared to similar non-concentric craters
- spectroscopic studies showing the interior composition to be very similar to the surrounding terrain.

Of the various hypotheses, igneous intrusion appears to be the most likely candidate, and spectral studies support this model. Dr. David Trang of the University of Hawaii Institute of Geophysics and Planetology describes the process: "The idea suggests that dikes have propagated into impact-induced fractures producing

FIG. 8.5 View of concentric crater Apollonius N located in Apollo Basin. (Credit: NASA)

a laccolith (a mass of igneous rock that intrudes between sedimentary beds and produces a dome-like bulging of the overlying strata). The laccolith grows, uplifting the crater floor. The inner torus is created whether because impact melt in the center of the crater is inhibiting uplift of the floor relative to the sides or inflation is followed by a large deflation or collapse" (Fig. 8.6).

The process is further described by Dr. Christian Wohler of Bielefeld University, Germany: "At the bottom of the transient cavity, however, the flexural rigidity of the overburdened weights per unit area were reduced to 30 % and 61 % of the original values, respectively. (Overburden may be used as a term to describe all soil and ancillary material above the bedrock horizon in a given area.) The thinned part of the overburdened was probably unable to resist the pressurized magma, which in turn may have lifted up the crater floor, thus leading to the shallow crater depth. In this line of thought, the inner depression of the concentric crater is a remnant of the original bowl-shaped floor."

Fig. 8.6 Oblique view of concentric crater Apollonius N located in Apollo Basin

Based on this, it is proposed that many concentric craters are polygenetic structures—impact craters colonized by lava that used crater breccia and fracture zones as conduits to the surface. According to Wood: "For the lava to form rings rather than ponds requires a higher viscosity or a lower extrusion rate than normal mare lavas. The magma may have been mare basalts erupted under unusual conditions, or mare basalt that differentiated within pockets, or in some cases, a non-mare magma type. The variations of concentric crater morphologies suggest a variety of styles of volcanic modification. The descriptions of non-mare style volcanic landforms and speculations of non-mare magmas complement similar conclusion for the material that forms some steep lunar domes."

There is much more to be learned about these fascinating craters by more detailed study of each one to further characterize their morphologies and from spectral analysis of their interiors and surroundings (Table 8.1).

Table 8.1 List of concentric craters

	Crater	Coordinates
1.	Archytas G	0.4 E 55.8 N
2.	Archytas G-b	0.48E 55.56 N
3.	Reaumur B CC	1.3E 4.5S
4.	Egede G	6.8E 51.9 N
5.	Pontanus E	13.3 E 25.2S
6.	Torricelli R CC	26.6E 6.4S
7.	Beaumonth P CC	29.6E 19.1S
8.	Fracastorius E	31.02 20.23.S
9.	Crozier H	49.4 E 14.1S
10.	Endymion SW CC	50.6 E 51.7 N
11.	Apollonius N	64.0E 4.7 N
12.	Legendre CC	68.2 E 27.5S
13.	Dubyago CC	69.0E 3.7 N
14.	Schubert N CC	74.0E 2.0 N
15.	Humboldt CC	83.2E 26.5S
16.	Hamilton CC	84.7E 44.2S
17.	Jeans CC	94.4E 53.1S
18.	Chamberlain CC	102.5E 58.8S
19.	Pasteur CC	104.9E 11.8S
20.	Jules Verne CC	143.76E 37.23S
21.	Jules Verne CC SE	144.1E 37.40S
22.	Geiger CC	159.0E 16.3A
23.	Vertregt K CC	172.6E 20.6S
24.	MacMillan	7.8 W 24.1 N
25.	Hesodius A	17.0 W 30.1 S
26.	Gambart J	18.2 W 0.7S
27.	Laplace E CC	21.2 W 50.0 N
28.	Fontennelle D	23.3 W 62.5 N
29.	Biancanus C CC	29.1 W 66.2S
30.	Marth	29.3 W 31.1 S
31.	La Condamine F CC	31.3 W 57.2 N
32.	Hainzel H	33.1 W 36.9 S
33.	Bouguer B CC	33.8 W 53.5 N
34.	Bouguer A CC	34.1 W 53.2 N
35.	J. Herschel F CC	34.6 W 34.79 W
36.	Gruithuisen M CC	42.01 W 37.15 N
37.	Gruithuisen CC	41.4 W 36.7 N
38.	Gruithuisen K	42.7 W 35.3 N
39.	Clausius E CC	46.8 W 35.3S
40.	Mersenius S CC	46.60 W 17.80S
41.	Louville DA	51.6 W 46.6 N
42.	Mersenius M	48.3 W 21.3S
43.	Damoiseau BA	59.0 W 8.3 S

(continued)

Table 8.1 (continued)

	Crater	Coordinates
44.	Damoiseau D	63.28 W 6.44S
45.	Lagrange T CC	62.0 W 32.82 S
46.	Lagrange T	62.4 W 32.9 S
47.	Markov CC	64.8 W 52.6 N
48.	Cruger CC	65.7 W 17.0 S
49.	Rocca	67 1 W 14.2 S
50.	Cavalerius E	69.9 W 7.7 N
51.	Lavoisier A CC	75.0 W 36.7 N
52.	Repsold A	77.5 W 51.9 N
53.	Struve CC	77.7 W 22.0 N
54.	Lavoisier CC	81.1 W 38.4 N
55.	Minkowski CC	143.0 W 56.6 S
56.	Apollo CC	154.2 W 31.0 S
57.	De Vries CC	172.7 W 20.6S

Tale of a Tail

The Moon has a tail. And at its longest, it can stretch across a quarter of a million miles—long enough to reach Earth.

The "tail" is a thin cloud of sodium atoms ejected into space from the lunar surface. They most likely were blasted into space by energy or charged particles from the Sun, or by the impacts of meteorites, which vaporize material when they hit the surface.

Some of the sodium atoms fall back to the Moon, but some of them get enough "kick" to climb thousands of miles high. They form a thin atmosphere that completely encircles the Moon (Fig. 8.7).

The atoms have an electric charge, so many of them are swept up by the solar wind—a million-mile-an-hour stream of charged particles from the Sun. This sculpts the sodium into a teardrop-shaped tail that streams behind the Moon. When the Moon is new—when it passes between Earth and the Sun—the end of the tail can touch Earth. The tail is too faint to be visible.

A Japanese satellite that's orbiting the Moon found that the atmosphere and tail grow thinner as the Moon moves from first quarter to last quarter. It also found that it's thickest during February—although no one can yet explain why.

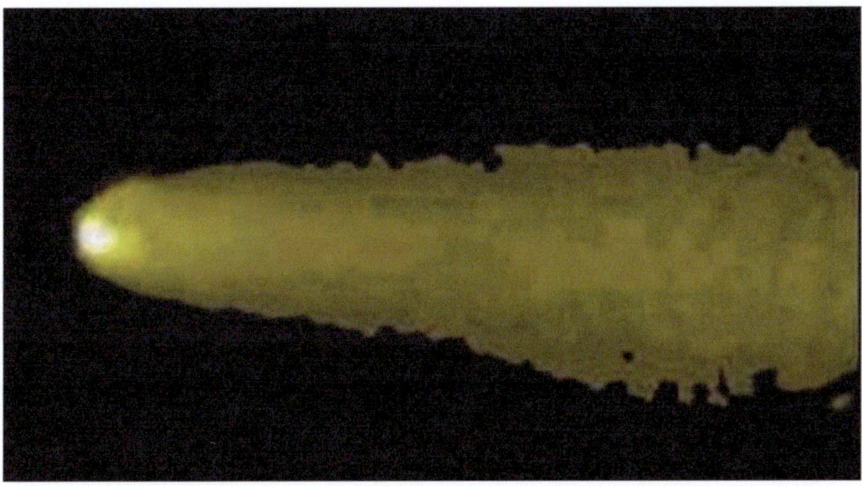

FIG. 8.7 The tail of sodium gas was seen to stretch for distances of at least 800,000 km (500,000 mi.) behind the Moon. It is very difficult to see. If the Moon's sodium tail were perhaps a thousand times brighter it would be bright enough for the human eye to see. It would appear as a glowing orange cloud dominating the night-time sky. (Credit: NASA)

The tail was discovered at the McDonald Observatory in Fort Davis, Texas, on nights of the Leonid meteor shower in November 1998 by a team from Boston University who used a special camera they designed that is sensitive to sodium wavelengths.

During the November observations, the BU team pointed their sensitive camera in the opposite direction from the Moon and recorded, just by chance, images of the tail in an otherwise moonless sky. "At the time of the Leonid meteor shower on November 17, 1998, the Moon was in a 'new' phase, impossible to see at its location between Earth and the Sun," said Dr. Steven Smith, research associate at the Center for Space Physics at BU. "Our team was operating on the night side of Earth, essentially looking away from the Sun and Moon, searching for meteor effects in our atmosphere." After one night of uneventful observations, on November 18 the imaging system detected a small patch of sodium emission in the dark skies above west Texas. "It grew to be larger and brighter on November 19, and then faded slightly on November 20," Smith said.

The BU team considered several theories that could explain these unusual features, ruling out a comet, the impact of Leonid meteors upon dust in the Solar System, and even possible instrumentation problems. Dr. Jody Wilson, research associate in the BU space physics group, suggested that the mysterious sodium gas might come from the Moon, and set out to model it using computer simulation and visualization techniques. "We found out that when the Moon is new, it takes 2 days or so for sodium atoms leaving the surface to reach the vicinity of Earth. They are pushed away from the Moon by the pressure of sunlight, and, as they sweep past us, Earth's gravity pulls on them, focusing them into a long narrow tail," Wilson explained.

"The pieces of the puzzle fit together rather well," said Michael Mendillo, professor of astronomy at BU. "While some of the Leonid meteors burned up in their streaks through Earth's atmosphere on the night of November 17—producing spectacular showers in some locations—others crashed into the Moon's dusty soil, liberating sodium gas. These atoms, speeding away from the Earth-Moon system, were then captured in photographs from our instrument in Texas several days later, looking down the length of the tail."

"If it were bright enough for the human eye to see, perhaps a thousand times brighter," added Jeffrey Baumgardner, Senior Research Associate at the University's Center for Space Physics, "it would be a glowing orange cloud dominating the nighttime, moonless sky."

The meteor shower temporarily tripled the mass of the tail by adding to the population of sodium atoms being outgassed by the Moon. The daily impact of small meteoroids and the like produces a constant "tail" on the Moon, but the Leonids had intensified it, thus making it more observable from Earth than normal.

Observing the Moon's sodium tail is a difficult task, especially when there is no meteor shower, but not impossible for amateur astronomers to attempt, notes Brian Cudnik in his book *Lunar Meteoroid Impacts and How to Observe Them*. "One needs to look at the time of new Moon in the direction of the nighttime sky directly opposite the Moon's location (or where the full Moon would be located on that day)," said Cudnik.

"It is only during this time that the sodium tail is very faintly visible (its density just above the Moon's surface is only a few atoms per cubic cm or less in Earth's lower atmosphere), under the most transparent and darkest of skies, and with the most sensitive detectors. This could possibly be done with a wide field, large aperture telescope equipped with a filter that can see the wavelength of light emitted from sodium atoms. This would need to be done from a dark sky site, well away from the ubiquitous sodium street lights found in towns and cities."

Try observing the tail during a meteor shower since meteor showers increase the density of the sodium tail from microscopic lunar impacts. Or try observing it near times of new Moon between April and November. Filtered images (Na D1 + D2) were made from the El Leoncito Observatory in Argentina using Boston University's All-Sky Imager (ASI). Four of the eight new Moon periods had clear skies, and images from those nights were calibrated to absolute brightness levels in Rayleighs (R) using standard stars. Results reveal variability in the brightness of the Moon's sodium tail that may be linked to enhanced contributions from meteor showers. Typically, observations begin approximately 1 h after sunset and end 1 h beforo sodium images taken nearest local midnight.

Observations of the lunar exosphere are uniquely difficult when the Moon is in the magnetotail, because the Moon is near full phase and the sunlight reflected off of its surface is at its brightest. This results in high scattered light both in Earth's atmosphere and in the sensitive instruments needed to observe the sodium emission. The solution is to conduct imaging observations during the totality phase of a lunar eclipse, when for approximately 1 h the Moon is within the umbra of Earth's shadow and sunlight scattered off of the surface is vastly reduced. Since Na is detected through its resonant scattering of sunlight, the reduced photon flux in Earth's umbra and penumbra limits detections close to the Moon.

Since the Apollo Moon program, scientists have known that the Moon has an atmosphere, but it is extremely thin. "It is one continuously being produced by evaporation of surface materials, and then continuously lost by escape or impact back onto the surface," said Mendillo.

Ten years ago, ground-based telescopes revealed that sodium gas formed part of the lunar atmosphere. "There are less than 50 atoms of sodium per cubic cm in the atmosphere just above the surface of the Moon," says Baumgardner. In contrast there are 10,000 million billion molecules per cubic cm in Earth's atmosphere at the surface.

For most practical purposes, the Moon is considered to be surrounded by vacuum. The elevated presence of atomic and molecular particles in its vicinity (compared to interplanetary medium), referred to as 'lunar atmosphere' for scientific objectives, is negligible in comparison with the gaseous envelope surrounding Earth and most planets of the Solar System—less than 100 trillionth (10^{-14}) of Earth's atmospheric density at sea level.

Mercury also has a glowing tail of sodium atoms. New measurements of Mercury's yellow-orange tail, which streams in the solar wind like the long tail of a kite, put it at more than 100 times the radius of the planet itself. The neutral sodium atoms that make up the 2.5 million km-long streamer are thought to be blasted off the surface by the Sun and micro-meteor impacts. These impart enough energy to launch the atoms into space.

Mercury and our Moon are not the only heavenly bodies with a sodium tail. A haze around Jupiter from the sodium blasted off of its tiny and hyper-volcanic Moon Io has also been seen. It's also been seen blowing from comets.

Because tails are associated with rocky bodies in our Solar System, they could someday help planet hunters identify rocky worlds orbiting other stars. It's a stepping stone to understanding other planets.

Tale of Another Tail

You might be zapped if you stand on the Moon when it's full. At that time, it develops a strong electric field near the surface as it swings through Earth's magnetic "tail," according to new observations from a Japanese probe.

Earth's magnetic field creates a protective bubble known as the magnetosphere, which surrounds the planet and shields us from solar wind—a rush of charged particles, or plasma, constantly streaming from the Sun (Fig. 8.8).

As the solar wind pushes on Earth's magnetic bubble, the planet's magnetosphere stretches, forming what's called the

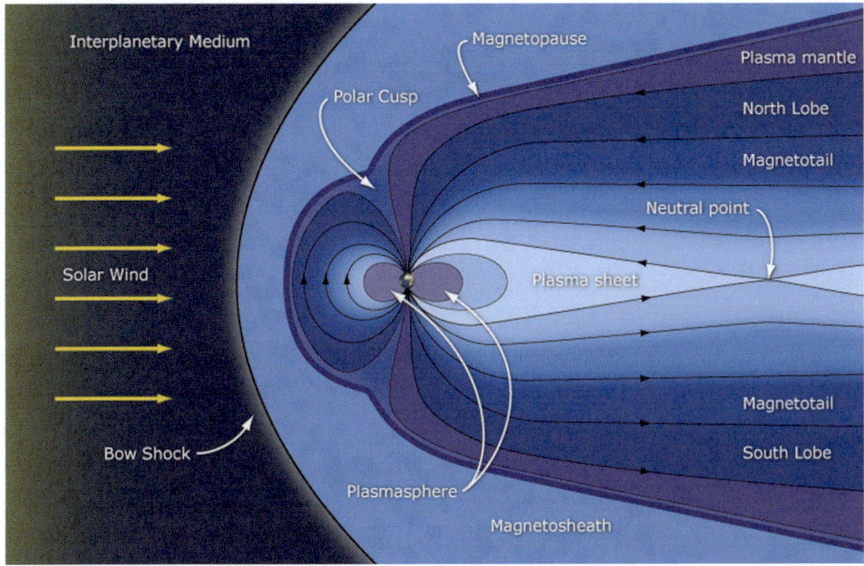

Fig. 8.8 Illustration of Earth's magnetic field, which looks a little like a sideways jellyfish. The jellyfish "tail" is known as the magnetotail, and it flows off to the right in this picture, away from the Sun on the "night side" of Earth. (Credit: NASA)

magnetotail. This tail reaches beyond the orbit of the Moon, and it's always pointed away from the Sun. Meanwhile, we see a full Moon when the lunar orb is on the opposite side of Earth from the Sun—and therefore within the magnetotail.

The magnetic tail is an extension of the same familiar magnetic field we experience when using a Boy Scout compass to find our way around Earth's surface. Our entire planet is enveloped in a bubble of magnetism, which springs from a molten dynamo in Earth's core. Out in space, the solar wind presses against this bubble and stretches it, creating a long magnetotail in the downwind direction.

A strong electric field near the surface was discovered by the Kayuga lunar orbiter, managed by the Japan Aerospace Exploration Agency (JAXA). JAXA launched the Kaguya probe in 2007. The craft orbited at a mere 62 miles (100 km) above the Moon's surface for 20 months, returning the first high-definition movies of the lunar landscape.

The orbiter was intentionally crashed into the Moon in June 2009. Scientists are now mining the data sent back by the craft's imagers and instruments, including the Magnetic field and Plasma experiment (MAP), which recorded that relatively high-energy electrons gyrating in the magnetic field are being absorbed by the lunar surface when the Moon is full. The result is a strong electric field that develops around the lunar surface around the same time as the full Moon, the Kaguya team reports in a paper published in the journal *Geophysical Research Letters*.

Based on Kaguya's data, the team says this relatively intense electric field can be found when the Moon passes through the region of the magnetosphere called the plasma sheet, which runs down the middle of Earth's magnetotail. Since the Moon has no global magnetic field of its own, its surface remains exposed to the trapped charged particles inside the plasma sheet.

"We are suggesting the presence of a relatively strong electric field around the Moon when it is inside Earth's magnetosphere, but the origin of this electric field remains a mystery," said study leader Yuki Harada, a graduate student at Kyoto University in Japan.

"We think it may be related to the properties of the plasma and magnetic field in Earth's magnetotail, and also the interaction between the Moon and surrounding plasmas."

The newly seen effect adds to what scientists know about the Moon's electrical activity. For example, previous observations have shown that the Moon gets slightly electrified by sunlight, regardless of whether the Moon is in Earth's magnetotail.

The day side of the Moon becomes positively charged, as solar radiation knocks electrons from the surface. Meanwhile, electrons build up on the night side of the Moon and give the surface a negative charge, Harada said.

The big question now is whether a strongly charged lunar surface poses risks to future robotic and human explorers "It is quite possible that electric fields induce a charge-up and subsequent discharge around a space vehicle, which could bring about serious damages to the human missions," Harada said.

The biggest hazards of an electrified Moon may be that the static charge can transport large amounts of the Moon's abrasive dust, which can damage sensitive lenses and electronics. The static buildup can also lead to unexpected electrical discharges.

NASA's Explorer series of probes and the Apollo missions were the first to reveal the perplexing lunar plasma environment, hinting that electrically driven dust may be a concern to robots and humans setting down on the Moon.

"We've been to the surface before and survived just fine, but we did have a number of problems with dust, among other things. And we happened to be there during very quiet plasma conditions," said Jasper Halekas, a plasma physicist at the University of California, Berkeley.

"Things might be very different during a solar storm, or during a passage through the plasma sheet, the region that was looked at in this study," he said.

"Certainly when you have big electric fields, you start to worry about damage to sensitive electronics, etc. And if those electric fields mobilize dust, that could become an additional problem."

But, he said, "the truth is that we don't really know yet what relevance these kinds of studies may have for exploration. Probably the only way we will ever know for sure is to go back to the surface."

Much of this is pure speculation. No one can say for sure what happens on the Moon when the magnetotail hits, because no one has been there at the crucial time. Apollo astronauts never landed on a full Moon, and they never experienced the magnetotail.

The best direct evidence comes from NASA's Lunar Prospector spacecraft, which orbited the Moon in 1998–1999 and monitored many magnetotail crossings. During some crossings, the spacecraft sensed big changes in the lunar night-side voltage, jumping "typically from –200 V to –1000 V," says Jasper Halekas of UC Berkeley, who has been studying the decade-old data.

"It is important to note," says Halekas, "that the plasma sheet (where all the electrons come from) is a very dynamic structure. The plasma sheet is in a constant state of motion, flapping up and down all the time. So as the Moon orbits through the magnetotail, the plasma sheet can sweep across it many times. Depending on how dynamic things are, we can encounter the plasma sheet many times during a single pass through the magnetotail, with encounters lasting anywhere from minutes to hours or even days."

"As a result, you can imagine how dynamic the charging environment on the Moon is. The Moon can be just sitting there in a quiet region of the magnetotail and then suddenly all this hot plasma goes sweeping by, causing the night-side potential to spike to a kilovolt. Then it drops back again just as quickly." The roller coaster of charge would be at its most dizzying during solar and geomagnetic storms. "That is a very dynamic time for the plasma sheet, and we need to study what happens then," he says.

Earth's magnetotail isn't the only source of plasma to charge the Moon. Solar wind can provide charged particles, too; indeed, most of the time, the solar wind is the primary source. But when the Moon enters the magnetotail, the solar wind is pushed back and the plasma sheet takes over. The plasma sheet is about ten times hotter than the solar wind, and that gives it more "punch" when it comes to altering the charge balance of the Moon's surface. Two-million-degree electrons in the plasma sheet race around like crazy, and many of them hit the Moon's surface. Solar wind electrons are relatively cool at only 140,000°, and fewer of them zip all the way down to the shadowed surface of the Moon's night-side.

Paved with Glass

Lunar explorations have revealed that much of the Moon's surface is covered with a glassy glaze, which indicates that the Moon's surface has been scorched by an unknown source of intense heat. As one scientist put it, the Moon is "paved with glass." One explanation forwarded was that an intense solar flare of awesome proportions scorched the Moon some 30,000 years or so ago. Scientists have remarked that the glassy glaze is not unlike that created by atomic weapons (the high radiation of the Moon should also be considered in light of this theory).

The astronomer who put forth this idea was Dr. Thomas Gold of Cornell University, a member of the U. S. National Academy of Sciences and a Fellow of the Royal Society. He speculated that in the not too distant past a violent solar flare heated up the surface of the Moon to the melting point. The same flare was also blamed by the renowned Carl Sagan for sweeping away the entire atmosphere of Mars.

After examining Moon rocks brought back by Apollo astronauts, Gold published a report in which he estimated that the flare had seared the Moon with radiation intensity equal to 100 suns for 10–100 s. Photographs taken by *Apollo 11* astronauts showed that this had caused the glass to liquefy and trickle down the side of many rocks.

In addition to glass coatings, close-up photos showed glass spheres and other particles outside nearby craters around the landing module.

Solar flares can unleash coronal mass ejections (CMEs), which are massive clouds of hot plasma and charged particles that blow outward from the surface of the Sun and stream into space. A strong CME can contain roughly a billion tons of plasma that tear through space at a blistering pace of up to a million miles per hour in a cloud many times the size of Earth.

Since the Moon has almost no atmosphere, Earth's natural satellite is vulnerable to the devastating effects of this space weathering. As a result, plasma from CMEs is able to easily penetrate the Moon's weak exosphere and blast the surface.

Between 1969 and 1972 six Apollo missions brought back 382 kg (842 lbs.) of lunar rocks, core samples, pebbles, sand, and dust from the lunar surface. Glass made up fully one-half of the Moon-soil sample brought back to Earth. About 5 % of the glass consisted of delicate globules and teardrops that show beautiful shades of brown, green, wine-red, and lemon.

The six spaceflights returned 2200 separate samples from six different exploration sites on the Moon. In addition, three automated Soviet spacecraft returned important samples totaling 300 g (approximately 3/4 lb.) from three other lunar sites. The lunar sample building at Johnson Space Center is the chief repository for the Apollo samples. This is where pristine lunar samples are prepared for shipment to scientists and educators around the world. Nearly 400 samples are distributed each year for research and teaching projects.

In addition to solar flares, intense meteorite bombardment and lava outpourings from volcanoes can also produce lunar glass. The various kinds of glasses include agglutinate particles and volcanic and impact spherules. The agglutinates form at the lunar surface by micrometeorite impacts that cause small-scale melting

Fig. **8.9** *Apollo 17* astronauts discovered an area of orange soil on the rim of Shorty Crater, in the Valley of Taurus-Littrow. (Credit: NASA)

that fuses adjacent materials together with tiny specks of metallic iron embedded in each dust particle's glassy shell. Glasses vary considerably from one site to another (Fig. 8.9).

The orange glass spheres and fragments from the *Apollo 17* Taurus-Littrow landing site are the finest particles ever brought back from the Moon. The particles range in size from 20 to 45 μm. They were discovered in the orange soil at Shorty Crater by scientist-astronaut Harrison J. Schmitt.

The orange particles, which are intermixed with black and black-speckled grains, are about the same size as the particles that compose silt on Earth. Chemical analysis of the orange soil material has shown the sample to be similar to some of the samples brought back from the *Apollo 11* (Sea of Tranquility) site several hundred miles to the southwest.

Like those samples, the one is rich in titanium (8 %) and iron oxide (22 %). But unlike the *Apollo 11* samples, the orange soil is

FIG. 8.10 **Moon marbles** These round spheres of green glass were collected from the lunar surface by *Apollo 15* astronauts Jim Irwin and Dave Scott from a location called Spur Crater. They are the result of volcanic activity on the Moon, quickly crystallized bits of molten material that was ejected from an ancient volcano 3.8 billion years ago. Just goes to show that the surface of the Moon holds a lot more than just gray rocks and dust! (Credit: NASA)

unexplainably rich in zinc. The orange soil is probably of volcanic origin and not the product of meteorite impact.

The *Apollo 17* soils contain a large volcanic pyroclastic contribution, most of which consists of glass spheres (Fig. 8.10).

The *Apollo 15* data are particularly instructive in that the site is on the mare surface but a short distance from the Apennine front. Both highland and mare rock compositions are, therefore, well represented. To add to the complexity of the *Apollo 15* site the soils also contain a modest proportion of green-glass particles.

These particles are basaltic in composition and are believed to be pyroclastic in origin. A total of eleven compositional types occur among the homogeneous glasses from the *Apollo 15* site, of which green glass particles are the most common.

Scientists have recently found water in some tiny beads of volcanic glass that Apollo astronauts collected on the Moon decades ago, raising new questions about how the Moon was born. For a long time, scientists have thought that the Moon is completely dry. They believed there couldn't be water on the Moon because of how it was formed. According to the "giant impact" theory, the Moon was born 4.5 billion years ago when an object the size of Mars came hurtling out of the void and smacked the young Earth. This impact melted both objects and scattered a cloud of debris around Earth. That debris became the Moon—and the new Moon was a hot ocean of magma. Researchers thought there was no way a Moon this hot could have retained any water.

But Erik Hauri, a geochemist at the Carnegie Institution's Department of Terrestrial Magnetism in Washington, D. C., got together with some colleagues to look for water in a special kind of material collected on the Moon by Apollo astronauts: tiny bits of glass created by volcanic eruptions on the early Moon.

"They're very tiny, the size of a period on a typical printed page," says Hauri, who explains that as volcanoes on the Moon erupted, tiny droplets cooled in the air, turning to glass before they even hit the ground. The glass beads come in different colors. Some were almost invisible and had to be picked out of gray Moon dust, but others could be seen by the astronauts as they walked around on the lunar surface.

A study about 10 years ago looked at the beads and saw hints of water. But the results also fell within the instrument's margin of error. So Hauri and his colleagues decided to look again using newer, more sensitive detection technology.

Some Moon experts were skeptical, saying there was no way the team would find any water molecules. "We got a lot of comments like, 'Oh, well, we already know that the Moon is dry, we've known this for 20 years,'" Hauri says.

But this time they did find water molecules, according to a report in the journal *Nature*. It wasn't a lot of water. The beads had only up to 46 ppm. But from this, the researchers estimate that the interior of the Moon once probably contained an amount of water equal to that of the Caribbean Sea.

This is a problem for the giant impact theory, says Hauri. "It's hard to imagine a scenario in which a giant impact melts, completely, the Moon, and at the same time allows it to hold onto its water," he says. "That's a really, really difficult knot to untie."

Others agree that the discovery raises questions. "It's obviously a very new result to say that the Moon might have had a significant amount of water right after it formed," says Ben Bussey, with the Johns Hopkins University Applied Physics Lab in Laurel, MD.

Scientists have seen hints of ice at the Moon's poles, in cold, shadowed craters. But because the Moon was assumed to be dry, researchers thought this ice had to come from comet impacts.

"With the cost of $25,000 for taking one pint of water to the Moon, it is essential that we develop processes of producing water from the materials on the Moon," said the study's lead author, Yang Liu, at the University of Tennessee at Knoxville. "This is paramount to human settlement of the Moon in the near future."

"This water would be of most value as rocket fuel—liquid hydrogen and liquid oxygen," Liu added. "Until the recent discovery of water in and on the Moon, this was going to be a very energy-intensive endeavor to separate these elements from the lunar rocks and soil. Now we have ready sources of water that can be consumed by plants and humans, but also broken up into its constituent elements—oxygen and hydrogen. Thus, we could use the Moon as a jump-board for missions to Mars and beyond."

It remained uncertain where all of this water might come from, although some apparently came from ice-rich comets. To find out more, scientists analyzed lunar surface dust, or regolith, that astronauts on the Apollo missions brought from the Moon. "Most samples actually come from an *Apollo 11* soil collected by Neil Armstrong," Liu noted.

Lunar regolith is created by meteoroids and charged particles constantly bombarding lunar rock. The researchers focused on grains of glass in the samples that were created in the heat of countless micrometeoroid impacts on the Moon. They reasoned this glass might have captured any water in the regolith before it cooled and solidified.

The investigators found that a large percentage of this glass contained traces of wetness—between 200 and 300 ppm of water and the molecule hydroxyl, which is much like water save that each of its molecules possesses just one hydrogen atom, not two.

To figure out where this water and hydroxyl originated from, the scientists looked at their hydrogen components. Hydrogen atoms come in a variety of isotopes, each with a different number of neutrons in their nuclei; regular hydrogen has no neutrons, while the isotope known as deuterium has one in each atomic nucleus.

The Sun is naturally low in deuterium because its nuclear activity rapidly consumes the isotope. All other objects in the Solar System possess relatively high levels of it, remnants of deuterium that existed in the nebula of gas and dust that gave birth to the Solar System.

The researchers found that the water and hydroxyl seen in the lunar glass were both low in deuterium. This suggests their hydrogen came from the Sun, probably blasted onto the Moon via winds of charged particles from the Sun, which continuously stream from the Sun at a rate of 2.2 billion lbs. (1 billion kg) per second. The Moon, lacking a significant atmosphere or magnetic field, slowly captures all the particles striking it. The hydrogen particles then bonded with oxygen bound in rocks on the lunar surface.

"The origin of surface water on the Moon was unclear," Liu said. "We provide robust evidence for a solar wind origin. This finding emphasizes the potential in finding such water on the surface of other similar airless bodies, such as Eros, Deimos, Vesta." The scientists detailed their findings in the journal *Nature Geoscience*.

Dr. Marek Zbik, a soil scientist from the Queensland University of Technology's Science and Engineering in Australia, also made an interesting discovery recently in the bubbles of glass from the Moon. Dr. Zbik took the lunar soil samples to Taiwan, where he could study the glass bubbles without breaking them using a new technique for studying nano-materials called synchrotron-based nano tomography to look at the particles. Nano tomography is a transmission X-ray microscope that enables 3D images of nano particles.

"We were really surprised at what we found," Dr. Zbik said. "Instead of gas or vapor inside the bubbles, which we would expect to find in such bubbles on Earth, the lunar glass bubbles were filled with a highly porous network of glassy-looking particles that span the bubbles' interior.

"It appears that the nano particles are formed inside bubbles of molten rocks when meteorites hit the lunar surface. Then they are released when the glass bubbles are pulverized by the consequent bombardment of meteorites on the Moon's surface."

"This continuous pulverizing of rocks on the lunar surface and constant mixing develop a type of soil which is unknown on Earth."

Zbik said nano particles behave according to the laws of quantum physics, which are completely different from so called 'normal' physics' laws. Because of this, materials containing nano particles behave strangely according to our current understanding. "Nano particles are so tiny, it is their size and not what they are made of that accounts for their exceptional properties."

"We don't understand a lot about quantum physics yet, but it could be that these nano particles, when liberated from their glass bubble, mix with the other soil constituents and give lunar soil its unusual properties."

"Lunar soil is electro-statically charged so it hovers above the surface; it is extremely chemically active; and it has low thermal conductivity (e.g., it can be 160° above the surface, but −40° two m below the surface). It is also very sticky and brittle such that its particles wear the surface off metal and glass."

Zbik said the Moon had no atmosphere to cushion the impact of meteorites, like Earth has. "When they hit the Moon there is a very violent reaction. Huge temperatures are generated that melts the rock. The pressure goes, and a vacuum is created. Bubbles occur in the molten glass rock like soft drink bubbles trying to escape the bottle. Our work now is to understand how those particles evolve from this process. It may also lead us to completely different way of manufacturing nano materials."

Zbik and his research team's study was published in the *International Scholarly Research Network Astronomy and Astrophysics.*

Dark Shadows

On the next sunny day, step outdoors and look inside your shadow. It's not very dark, is it? Grass, sidewalk, toes—whatever's in there, you can see quite well.

Your shadow's inner light comes from the sky. Molecules in Earth's atmosphere scatter sunlight (blue more than red) in all directions, and some of that light lands in your shadow. Look at your shadowed footprints on fresh sunlit snow: they are blue!

Fig. 8.11 Blinding sunshine, dark shadows and the lunar lander Antares. (Credit: NASA)

Without the blue sky, your shadow would be eerily dark, like a piece of night following you around. Weird. Yet that's exactly how it is on the Moon (Fig. 8.11).

Why are shadows on the Moon so dark? On Earth, air scatters light and allows objects not in direct sunlight to still be well-lit. This is an effect called Rayleigh scattering, named for the British Nobel-winning physicist Lord Rayleigh (John William Strutt). Rayleigh scattering is the reason why the sky is blue, and (for the most part) why you can still read a magazine perfectly well under an umbrella at the beach.

On the Moon there is no air, no Rayleigh scattering. So shadows are very dark and, where sunlight hits, very bright. Shadowed areas are dramatically murky, yet there's still *some* light bouncing around in there—due to reflected light from the lunar surface itself.

To visualize the experience of Apollo astronauts, imagine the sky turning completely and utterly black while the Sun continues to glare. Your silhouette darkens, telling you "you're not on Earth anymore."

That's what Neil Armstrong and Buzz Aldrin quickly discovered when they first walked the landscape of the Moon. Everything the Sun didn't shine directly on was pitch black. Once their foot stepped in a shadow, they could not see it anymore despite the fact that the Sun was blazing in the sky.

Shadows were one of the first things *Apollo 11* astronaut Neil Armstrong mentioned when he stepped onto the surface of the Moon. "It's quite dark here in the shadow [of the lunar module] and a little hard for me to see that I have good footing," he radioed to Earth.

The Eagle had touched down on the Sea of Tranquility with its external equipment locker, a stowage compartment called MESA, in the shadow of the spacecraft. Although the Sun was blazing down around them, Armstrong and Buzz Aldrin had to work in the dark to deploy their TV camera and various geology tools.

Although they soon found they could adjust to the shadows, the constant contrast between dark shadowy areas and sunny ones remained a challenge. Things got even stranger when they noticed that some of the shadows—namely, their own—had halos. They later learned that their eerie experience was caused by the opposition effect, a phenomenon that makes certain dark, shadowed areas appear surrounded by a bright aureole when they're viewed in a certain angle to the Sun.

"It is very easy to see in the shadows after you adapt for a while," noted Armstrong. But, added Aldrin, "continually moving back and forth from sunlight to shadow should be avoided because it's going to cost you some time in perception ability."

Truly, Moon shadows aren't absolutely black. Sunlight reflected from the Moon's gently rounded terrain provides some feeble illumination, as does Earth itself, which is a secondary source of light in lunar skies.

The shadows on the Moon caused mischief on many Apollo missions. Some astronauts found their maintenance tasks impossible to perform because their own hands blocked out what they were doing. Others thought they were standing on a steep slope because the deep shadows seemed like a cavern.

Given plenty of time to adapt, an astronaut could see almost anywhere. *Almost.* Consider the experience of *Apollo 14* astronauts Al Shepard and Ed Mitchell. They had just landed at Fra Mauro and were busily unloading the lunar module. Out came the ALSEP, a group of experiments bolted to a pallet. Items on the pallet were held down by "Boyd bolts," each bolt recessed in a sleeve used to guide the Universal Handling Tool, a sort of astronaut's wrench. Shepard would insert the tool and give it a twist to release the bolt—simple, except that the sleeves quickly filled with moondust. The sleeve made its own little shadow, so "Al was looking at it, trying to see inside. And he couldn't get the tool in and couldn't get it released—and he couldn't see it," recalls Mitchell.

"Remember," adds Mitchell, "on the lunar surface there's no air to refract light—so unless you've got direct sunlight, there's no way in hell you can see anything. It was just pitch black. That's an amazing phenomenon on an airless planet."

(Eventually they solved the problem by turning the entire pallet upside down and shaking loose the moondust. Some of the Boyd bolts, loosened better than they thought, rained down as well.)

Tiny little shadows in unexpected places would vex astronauts throughout the Apollo program—a bolt here, a recessed oxygen gauge there. These were minor workaday nuisances, mostly, but astronauts were frustrated by the minutes lost from their schedule.

Shadows could also be mischievous. *Apollo 12* astronauts Pete Conrad and Al Bean landed in the Ocean of Storms only about 600 yards from *Surveyor 3*, a robotic spacecraft sent by NASA to the Moon 3 years earlier. A key goal of the *Apollo 12* mission was to visit *Surveyor 3*, to retrieve its TV camera, and to see how well the craft had endured the harsh lunar environment. *Surveyor 3* sat in a shallow crater where Conrad and Bean could easily get at it—or so mission planners thought.

The astronauts could see *Surveyor 3* from their lunar module Intrepid. "I remember the first time I looked at it," recalls Bean. "I thought it was on a slope of 40°. How are we going to get down there? I remember us talking about it in the cabin, about having to use ropes." But "it turned out [the ground] was real flat," rejoined Conrad.

What happened? When Conrad and Bean landed, the Sun was low in the sky. The top of *Surveyor 3* was sunlit, while the bottom was in deep darkness. "I was fooled," says Bean, "because, on Earth, if something is sunny on one side and very dark on the other, it has to be on a tremendous slope." In the end, they walked down a gentle 10° incline to *Surveyor 3*—no ropes required.

Bean and Conrad were tricked not once but twice. "A real interesting thing has happened to the solar wind collector," Bean radioed to Earth from inside the Intrepid. "When I left it yesterday it was just a flat sheet of foil, but as I look out there now, it has folded back around the pole that's holding it. Looks almost like a sail in the wind. It's sort of bulging in front and bent back on the sides. It's real crazy."

"You've got a fairly strong solar wind, I suspect," joked Ed Gibson by radio from mission control in Houston. Gibson knew, as did Bean, that the solar wind is too diaphanous to flutter a sail.

Indeed, when Bean went out to the collector to take pictures, he discovered that the foil looked absolutely normal. "I guess it was sort of an optical illusion from inside the spacecraft," he reported. "The thing that fools you," he explained later, "is the relative lightness and darkness of shadows on the object. [We] looked at the solar wind collector, and we think 'Man, that thing is really bent around the pole.' But we go out there and see that it's not" (Fig. 8.12).

A final twist: When astronauts looked at the shadows of their own heads, they saw a strange glow. Buzz Aldrin was the first to report "...[there's] a halo around the shadow of my helmet." Armstrong had one, too.

This is the "opposition effect." Atmospheric optics expert Les Cowley explains: "Grains of moondust stick together to make fluffy tower-like structures, called 'fairy castles,' which cast deep shadows." Some researchers believe that the lunar surface is studded with these microscopic towers. "Directly opposite the Sun," he continues, "each dust tower hides its own shadow and so that area looks brighter by contrast with the surroundings."

Sounds simple? It's not. Other factors add to the glare. The lunar surface is sprinkled with glassy spherules (think of them as lunar dew drops) and crystalline minerals, which can reflect sunlight backwards. And then there's "coherent backscatter"—specks

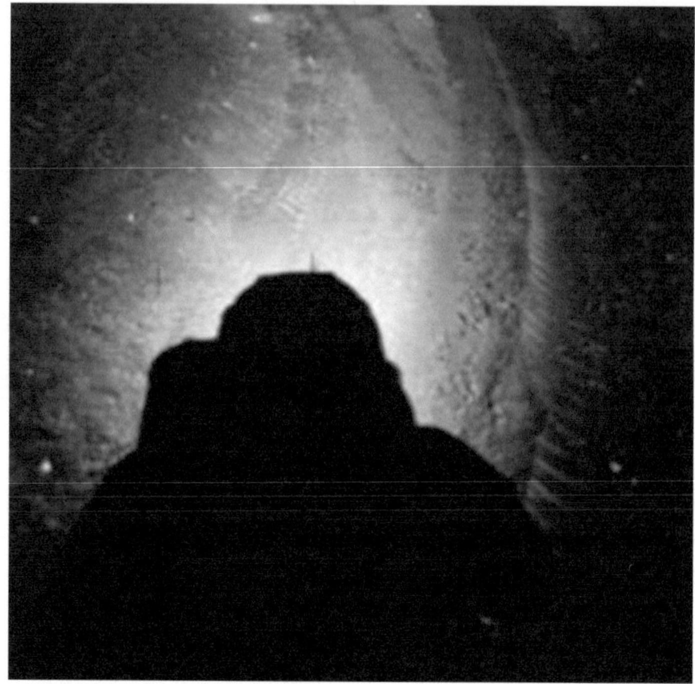

FIG. 8.12 When astronauts looked at the shadows of their own heads, they saw a strange glow. Buzz Aldrin was the first to report "...[there's] a halo around the shadow of my helmet." Armstrong had one, too

of moondust smaller than the wavelength of light diffract sunlight, scattering rays back toward the Sun. "No one knows which factor is most important," says Cowley.

We can experience the opposite effect here on Earth, for example, looking away from the Sun into a field of tall dewy grass. The halo is there, but our bright blue sky tends to diminish the contrast. For full effect, you've got to go to the Moon.

Luminous halos; mind-bending shadows; levitating moondust. Apollo astronauts had discovered a strange world indeed.

Stranger Than Fiction

A strange storm stirs the surface of the Moon every morning when the Sun first peeks over the dusty soil of the Moon after 2 weeks of frigid lunar night.

The next time you see the Moon, trace your finger along the terminator, the dividing line between lunar night and day. That's where the storm is. It's a long and skinny dust storm, stretching all the way from the north pole to the south pole, swirling across the surface, following the terminator as sunrise ceaselessly sweeps around the Moon.

Never heard of it? Few have. But scientists are increasingly confident that the storm is real.

The evidence comes from an old Apollo experiment called LEAM, short for Lunar Ejecta and Meteorites. *Apollo 17* astronauts installed LEAM on the Moon in 1972, explains Timothy Stubbs of the Solar System Exploration Division at NASA's Goddard Space Flight Center. "It was designed to look for dust kicked up by small meteoroids hitting the Moon's surface" (Fig. 8.13).

Fɪɢ. **8.13** The lunar terminator is the division between the illuminated and dark parts of Earth's Moon. (Credit: NASA)

Billions of years ago, meteoroids hit the Moon almost constantly, pulverizing rocks and coating the Moon's surface with their dusty debris. Indeed, this is the reason why the Moon is so dusty. Today these impacts happen less often, but they still happen.

Apollo-era scientists wanted to know, how much dust is ejected by daily impacts? And what are the properties of that dust? LEAM was to answer these questions using three sensors that could record the speed, energy, and direction of tiny particles—one each pointing up, east, and west.

LEAM's four-decade-old data are so intriguing, they're now being reexamined by several independent groups of NASA and university scientists. Gary Olhoeft, professor of geophysics at the Colorado School of Mines in Golden, is one of them: "To everyone's surprise," says Olhoeft, "LEAM saw a large number of particles every morning, mostly coming from the east or west—rather than above or below—and mostly slower than speeds expected for lunar ejecta."

What could cause this? Stubbs has an idea: "The day side of the Moon is positively charged; the night side is negatively charged." At the interface between night and day, he explains, "electrostatically charged dust would be pushed across the terminator sideways," by horizontal electric fields.

Even more surprising, Olhoeft continues, a few hours after every lunar sunrise, the experiment's temperature rocketed so high—near that of boiling water—that the LEAM had to be turned off because it was overheating.

Those strange observations could mean that "electrically-charged moondust was sticking to LEAM, darkening its surface so the experiment package absorbed rather than reflected sunlight," speculates Olhoeft.

But nobody knows for sure. LEAM operated for a very short time; only 620 h of data were gathered during the icy lunar night and a mere 150 h of data from the blazing lunar day before its sensors were turned off and the Apollo program ended.

Astronauts may have seen the storms, too. While orbiting the Moon, the crews of *Apollo 8, 10, 12,* and *17* sketched "bands" or "twilight rays" where sunlight was apparently filtering through dust above the Moon's surface. This happened before each lunar sunrise and just after each lunar sunset. NASA's Surveyor spacecraft also photographed twilight "horizon glows," much like what the astronauts saw.

It's even possible that these storms have been spotted from Earth. For centuries, there have been reports of strange glowing lights on the Moon, known as lunar transient phenomena, or LTPs, discussed earlier in this chapter. Some LTPs have been observed as momentary flashes—now generally accepted to be visible evidence of meteoroids impacting the lunar surface. But others have appeared as amorphous reddish or whitish glows or even as dusky hazy regions that change shape or disappear over seconds or minutes. Early explanations, never satisfactory, ranged from volcanic gases to observers' overactive imaginations (including visiting extraterrestrials).

Now a new scientific explanation is gaining traction. "It may be that LTPs are caused by sunlight reflecting off rising plumes of electrostatically lofted lunar dust," Olhoeft suggests.

All this matters to NASA and the commercial space industry because, by 2020 or so, astronauts expect to return to the Moon. Unlike Apollo astronauts, who never experienced lunar sunrise, the next explorers are going to establish a permanent outpost. They'll be there in the morning when the storm sweeps by.

The wall of dust, if it exists, might be diaphanous, invisible, harmless. Or it could be a real problem, clogging spacesuits, coating surfaces, and causing hardware to overheat.

Which will it be? Says Stubbs, "we've still got a lot to learn about the Moon."

Back in 1956, Hal Clement wrote a short story called "Dust Rag" published in the magazine *Astounding Science Fiction*, about two astronauts descending into a crater on the Moon to investigate a mysterious haze dimming stars near the lunar horizon. After discarding a wild guess that they were seeing traces of a lunar atmosphere—"gases don't behave that way"—they figured it had to be dust somehow suspended above the ground. In a conversation remarkable for its scientific prescience, one of the astronauts explains:

"The [Moon's] surface material is one of the lousiest imaginable electrical conductors, so the dust normally on the surface picks up and keeps a charge. And what, dear student, happens to particles carrying like electrical charges?"

"They are repelled from each other."

"Head of the class. And if a 100-km circle with a rim a couple of [km] high is charged all over, what happens to the dust lying on it?"

The answer, given only by narrative description, is that electrostatic charging caused the dust to levitate.

Well, guess what? Writer Clement was even more correct than he knew. It appears lunar dust does levitate above the Moon's surface because of electrostatic charging. And the first evidence came almost the way Clement had described.

In the early 1960s, before *Apollo 11*, several early Surveyor spacecraft that soft-landed on the Moon returned photographs showing an unmistakable twilight glow low over the lunar horizon persisting after the Sun had set. Moreover, the distant horizon between land and sky did not look razor-sharp, as would have been expected in a vacuum where there was no atmospheric haze.

But most amazing of all, *Apollo 17* astronauts orbiting the Moon in 1972 repeatedly saw and sketched what they variously called "bands," "streamers," or "twilight rays" for about 10 s before lunar sunrise or lunar sunset. Such rays were also reported by astronauts aboard *Apollo 8, 10,* and *15*.

Here on Earth we see something similar: crepuscular rays. These are shafts of light and shadow cast by mountain ridges at sunrise or sunset. We see the shafts when they pass through dusty air. Perhaps the Moon's "twilight rays" are caused, likewise, by mountain shadows passing through levitating moondust. Many planetary scientists in the 1970s thought so, and some of them wrote papers to that effect.

However, without an atmosphere, how could dust hover far above the Moon's surface? Even if temporarily kicked up by, say, a meteorite impact, wouldn't dust particles rapidly settle back onto the ground?

Well, no—at least not according to the "dynamic fountain model" for lunar dust recently proposed by Timothy J. Stubbs, Richard R. Vondrak, and William M. Farrell of the Laboratory for Extraterrestrial Physics at NASA's Goddard Space Flight Center.

"The Moon seems to have a tenuous atmosphere of moving dust particles," Stubbs explains. "We use the word 'fountain' to evoke the idea of a drinking fountain: the arc of water coming out of the spout looks static, but we know the water molecules are in motion." In the same way, individual bits of moondust are constantly leaping up from and falling back to the Moon's surface, giving rise to a "dust atmosphere" that looks static but is composed of dust particles in constant motion.

Rub an inflated balloon on your hair, and then hold the balloon a few inches away. Your hair will rise of its own accord to reach out toward the balloon. Rubbing the balloon removes some of the electrons from your hair, leaving your hair with a net positive charge. Your positively charged hair is attracted to the negatively charged balloon.

Now watch what happens when you hold the balloon far away. This is key. Your individual hairs spread out from one another and do not immediately fall back to lie flat on your head. What's happened? When the balloon was removed, each positively charged hair repels its positively charged neighbor and some of your hair remains suspended—just like dust on the Moon.

On the Moon, there is no rubbing. The dust is electrostatically charged by the Sun in two different ways—by sunlight itself and by charged particles flowing out from the Sun (the solar wind).

On the daylight side of the Moon, solar ultraviolet and X-ray radiation is so energetic that it knocks electrons out of atoms and molecules in the lunar soil. Positive charges build up until the tiniest particles of lunar dust (measuring 1 μm and smaller) are repelled from the surface and lofted anywhere from m to km high, with the smallest particles reaching the highest altitudes, Stubbs explains. Eventually they fall back toward the surface, where the process is repeated over and over again.

If that's what happens on the day side of the Moon, the natural question then becomes, what happens on the night side? The dust there, Stubbs believes, is negatively charged. This charge comes from electrons in the solar wind, which flow around the Moon onto the night side. Indeed, the fountain model suggests that the night side would charge up to higher voltages than the day side, possibly launching dust particles to higher velocities and altitudes.

Day side: positive. Night side: negative. What, then, happens at the Moon's terminator—the moving line of sunrise or sunset between day and night?

There could be "significant horizontal electrical fields forming between the day and night areas, so there might be horizontal dust transport," Stubbs speculates. "Dust would get sucked across the terminator sideways." Because the biggest flows would involve microscopic particles too small to see with the naked eye, an astronaut would not notice dust speeding past. Still, if he or she were on the Moon's dark side and alert for lunar sunrise, the astronaut "might see a weird, shifting glow extending along the horizon, almost like a dancing curtain of light." Such a display might resemble pale auroras on Earth.

Stubbs and his colleagues are now hard at work on a host of fascinating questions. For example, there are deep craters at the lunar poles that never see sunlight. Would these craters have a strong surplus of negative charge? Astronauts need to know, because in the years ahead NASA and commercial ventures plan to send people back to the Moon, and deep dark craters are places where they might find pockets of frozen water—a crucial resource for any colony. Will they also encounter swarms of electric dust?

It's not science fiction any more.

Sparks Fly

Electric sparks may help break down lunar dirt, suggesting the Moon may be significantly more active than previously thought. In fact, this kind of sparking might take place throughout the Solar System, from Mercury to the satellites of Jupiter and Saturn, researchers note.

The giant craters that pockmark the face of the Moon bear witness to a violent past full of cosmic impacts. Meteorites continue to bombard the Moon, breaking down or "weathering" its soil. Indeed, the Moon is much harder hit than Earth, whose thick atmosphere protects the planet from many incoming space rocks.

The Moon is also constantly hit by high-energy electrically charged particles. For instance, electrons and protons often burst from the Sun, and more energetic particles known as cosmic rays also regularly strike from elsewhere in the Milky Way Galaxy.

Previous studies only investigated how these high-energy particles electrically charge the uppermost layer of the lunar surface. Lead study author Andrew Jordan, a space physicist at the University of New Hampshire, and his colleagues explored whether high-energy electrically charged particles might also penetrate deeper into the Moon.

Lunar soil is an electrical insulator, meaning it has extremely low conductivity. Any electrically charged particles that do penetrate deep into it can thus get trapped inside and accumulate.

For decades, scientists have known that this phenomenon, known as deep dielectric charging, can occur in spacecraft. If too much charge accumulates, the resulting electrical sparks can damage electronics.

The researchers focused on craters near the lunar poles that host permanently shadowed regions. These dark areas can reach temperatures as low as—400 °F (–240 °C). The colder lunar soil gets, the more electrically insulating it becomes.

To estimate how strong electric fields can get in the permanently shadowed areas of the Moon, the scientists developed a computer model using data from the Cosmic Ray Telescope for the Effects of Radiation (CRaTER) instrument aboard NASA's LRO and from the Electron, Proton, and Alpha Monitor (EPAM) on NASA's Advanced Composition Explorer spacecraft.

The computer model suggested that powerful electric fields can build up in the permanently shadowed regions of the Moon, leading to sparking that vaporizes tiny channels in lunar soil.

"The lunar community generally thinks that the main ongoing process of fragmenting soil grains is meteoritic impacts," said Jordan. "Our research suggests that breakdown weathering may also contribute to this fragmenting, at least within permanently shadowed regions on the Moon."

The Lyman Alpha Mapping Project (LAMP) instrument on the LRO "has detected unexpectedly dark soil within many permanently shadowed regions on the Moon," Jordan noted. "Some scientists think the soil appears dark because it is more porous or fluffier than soil outside those regions. This fluffy soil has been a mystery, since it requires the grains to be very small. Breakdown weathering seems to be a process that can produce such small grains. We'll have to see."

The most important implication of these findings "is that permanently shadowed regions of the Moon really are a special place," Jordan said. "They are so cold that they experience phenomena that warmer regions would not."

In addition, "any airless body that both has such cold regions and is also bombarded by enough energetic charged particles could experience breakdown weathering," Jordan said.

"For example, the cold, permanently shadowed regions of Mercury are exposed to even more solar energetic particles than the Moon," he added. "This work may also apply to some of the Moons inside the radiation belts of Jupiter and Saturn. If so, then breakdown weathering may have affected the physical and even chemical properties of the soil on these bodies. There are some fun possibilities, but only further research will be able to tell."

Intriguingly, the permanently shadowed regions of the Moon are home to water ice. "As of now, we don't know whether breakdown weathering may affect ice, but we'll be looking into that," Jordan said.

To confirm whether sparking happens on the Moon and elsewhere, Jordan noted that "sparks give off electromagnetic radiation at many wavelengths, including visible and radio waves, so in principle they are detectable. Detecting them during a large solar energetic particle event may require a combination of ground- and space-based observations."

In addition, "having shown that breakdown weathering is a possibility, we'd like to quantify better how it might affect the soil," Jordan said. "We're working to find out how much of the soil in permanently shadowed regions may have been affected and compare that to what meteorites do."

The scientists detailed their findings in the *Journal of Geophysical Research-Planets*.

Mystery of the Moon Trees

Scattered around our planet are hundreds of creatures that have been to the Moon and back again. None of them is human. They outnumber active astronauts 3:1. And most are missing.

They're trees. "Moon Trees."

NASA scientist Dave Williams has found over 40 of them, and he's looking for more. "They were just seeds when they left Earth

Fig. 8.14 The Moon Trees are a living monument to our first visits to the Moon. (Credit: NASA)

in 1971 onboard *Apollo 14*," explains Williams. "Now they're fully grown. They look like ordinary trees—but they're special because they've been to the Moon." Moon trees can be found all over the world. They are usually spotted by accident because no one kept track of where most of them were planted. So far, Williams has located 44 of the 450 or so that were planted in 1976 (Fig. 8.14).

How they got here and back is a curious tale.

It begins in 1953, when Stuart Roosa parachuted into an Oregon forest fire. He had just taken a summer job as a U. S. Forest Service "smoke jumper," parachuting into wildfires in order to put them out. It was probably adventure that first attracted Roosa to the job, but he soon grew to love the forests, too. "My father had an affinity for the outdoors," recalls Air Force Lt. Col. Jack Roosa, Stuart's son. "He often reminisced about the tall Ponderosa pine trees from his smoke-jumping days."

Thirteen years later, NASA invited Roosa, who had since become an air force test pilot, to join the astronaut program. He accepted. Roosa, Ed Mitchell, and Al Shepard eventually formed the prime crew for *Apollo 14*, slated for launch in 1971.

"Each Apollo astronaut was allowed to take a small number of personal items to the Moon," continued Jack. Their PPKs, or Personal Preference Kits, were often filled with trinkets—coins, stamps or mission patches. Al Shepard took golf balls. On *Gemini 3*, John Young brought a corned beef sandwich. "My father chose trees," says Jack. "It was his way of paying tribute to the U. S. Forest Service."

The Forest Service was delighted.

"It was part science, part publicity stunt," laughs Stan Krugman, who was the U. S. Forest Service's staff director for forest genetics research in 1971. "The scientists wanted to find out what would happen to these seeds if they took a ride to the Moon. Would they sprout? Would the trees look normal?" In those days biologists had done few experiments in space; this would be one of the first. "We also wanted to give them away as part of the Bicentennial celebration in 1976."

Krugman himself selected the varieties: redwood, Loblolly pine, sycamore, Douglas fir and sweetgum. "I picked redwoods because they were well-known, and the others because they would grow well in many parts of the United States," he explained. "The seeds came from two Forest Service genetics institutes. In most cases we knew their parents (a key requirement for any post-flight genetic studies)."

On January 31, 1971, *Apollo 14* blasted off. Only Shepard and Mitchell actually walked on Moon. On February 5 they landed the lunar module Antares in Fra Mauro—a hilly area where Shepard famously launched his golf balls using a geology tool as a make-shift driver. Roosa remained in orbit as pilot of the mission's command module Kitty Hawk. Inside his PPK was a metal cylinder, 6 in. long and 3 in. wide, filled with seeds. Together they circled the Moon 34 times.

Apollo 14 was a success. Scientists were delighted with the mission's geology experiments, and they were eager to study the 43 kg of Moon rocks collected by Shepard and Mitchell. Krugman was just as eager to study the seeds.

"We had a bit of a scare," Krugman recalls. During decontamination procedures, the seed canister was exposed to vacuum, and it burst. The seeds were scattered and traumatized. "We weren't sure if they were still viable," he says. Working by hand, Krugman carefully separated the seeds by species and sent them to Forest Service labs in Mississippi and California. Despite the accident, nearly all of them germinated. "We had [hundreds of] seedlings that had been to the Moon!" Thirty-one years later, Krugman still sounded excited.

During the years that followed, the trees thrived as scientists watched. "The trees grew normally," he continued. "They reproduced with Earth trees and their offspring, called half-Moon trees, were normal, too." (He notes, however, that DNA analysis wasn't routinely done in the early '1970s, and so the Moon trees weren't tested in that way. There might be subtle differences yet to be discovered.)

Finally, in 1975, they were ready to leave the lab. "That's when things got out of hand," he says.

Everyone wanted a Moon tree. In 1975 and '1976, trees were sent to the White House, to Independence Square in Philadelphia, to Valley Forge. "One tree went to the Emperor of Japan. Senators wanted trees to dedicate buildings. We even did some plantings in New Orleans because the mayor there, Mayor Moon, wanted some," says Krugman. There were so many requests that "we had to produce additional seedlings from rooted cuttings of the original trees."

No one kept systematic records, notes Williams. That's why the whereabouts of the trees today are mostly unknown. Many ended up in out-of-the-way places. Williams' website encourages people to contact him with possible Moon tree sightings.

"I really like the fact that these trees can be anywhere, and I'm sure there are a lot out there in local parks, small college campuses and next to buildings that people are walking by every day without realizing it," Williams said. Most of the Moon trees that have been reported have a marker of some sort; without identification, there's no difference between them and standard-issue Earth trees. Williams didn't even know that there was a Moon tree at the Goddard Space Flight Center, where he works, until someone asked him why it wasn't listed on his website. "That was a little embarrassing," Williams said.

One of them went to a Girl Scout camp in Cannelton, Indiana, where third grade teacher Joan Goble found it in 1996. (She knew it was a Moon Tree because a sign said so. Most Moon trees were planted with ceremony; there's usually a sign or plaque nearby that identifies them.) "My students love it," she says. "It looks like an ordinary tree, but they feel it's special anyway because of its trip to the Moon." Jack Roosa has since become a pen pal of Goble's class, encouraging the students to explore and learn as his father did.

When contacted for more information about Moon trees, "I was clueless," Williams admits. Like many people who were young in the 1970s, Williams had never heard of such trees, but he soon became an enthusiast. "I found one Moon tree right here at Goddard near my office," he laughs. "I had no idea it was there."

Often that's how they're encountered—by accident. Williams now maintains a website listing all known Moon trees. If you stumble across one, contact Dave. He'll investigate the find and add it to the collection if it's authentic. His website and e-mail address is located at http://nssdc.gsfc.nasa.gov/planetary/lunar/moon_tree.html.

Moon trees are long-lived, adds Krugman. The redwoods could last thousands of years, and the pines have a life expectancy of centuries. Indeed, they've already outlived Stuart Roosa and Al Shepard—two of the humans who took them to the Moon.

Says Jack, "I think my father always knew that these trees would serve as a long-lasting, living reminder of mankind's greatest achievement—the manned missions to the Moon." Of course, if humans don't return soon, Moon trees could become the only living things on our planet that have been to the Moon. That's probably not what Stuart had in mind.

Jack, however, is optimistic: "These trees will be here 100 years from now," he says. "By then I believe we'll be planting Mars trees right beside them."

The Moon Is a Harsh Mistress

Poor Pluto. First it gets kicked out of the planet club, now it's not even the coldest known place in the Solar System. New data from NASA's LRO suggests that permanently shadowed craters at the

Moon's South Pole might be colder even than Pluto and the other objects in the Solar System's furthest reaches.

In its first set of measurements, LRO's Diviner Lunar Radiometer Experiment, which is conducting the first global survey of the temperature of the lunar surface, found that craters along the lunar south pole that have areas permanently shielded from the Sun's light (and suspected to harbor deposits of water-ice) have extremely cold temperatures.

"Diviner has recorded minimum daytime brightness temperatures in portions of these craters of less than –397 °F," said David Paige, Diviner's principal investigator and a UCLA professor of planetary science. "These super-cold brightness temperatures are, to our knowledge, among the lowest that have been measured anywhere in the Solar System, including the surface of Pluto." These temperatures are well below the freezing points of nitrogen and oxygen on Earth, and only a dozen degrees above the freezing point of hydrogen.

The measurements were made in October, when the Moon's North Pole was at mid-winter. Because the Moon is tilted a mere 1.54° on its axis (compared with 23.5° for Earth), its seasons are not very pronounced. But the poles have small regions where the Sun doesn't rise for around 6 months. It was there that the LRO's instruments measured the temperature.

Although it may seem odd that the Moon, which is much closer to the Sun, could be colder than Pluto, it's not at all unexpected, one planetary scientist said. In fact, the poles of Mercury may be even colder.

"The key point is not their distance from the Sun but the fact that there are regions at the poles of the Moon and Mercury that never see the Sun, and so never get heated by sunlight," said Alan Boss, of the Carnegie Institution in Washington, D. C.

"The only heat they receive is from the underlying rock, but that means only interior heat leftover from their formation or their internal radioactive decays, but in any case the local rocks are still cold because they, too, are free to radiate out to –263 °C space without getting any heat back from the Sun," Boss said.

The news of these frigid temperatures bolsters the idea that these craters could harbor water-ice, which would be a boon to any future Moon bases, which could melt the water and use it for drinking, or extract hydrogen for fuel.

The ultra-low temperatures of these craters are the opposite of those at the lunar equator, which are hotter than the boiling point of water at noontime. Lunar surface temperatures are expected to change with the seasons, and Diviner continued to monitor and map them throughout LRO's 1-year mission.

Temperatures were recorded at the southwestern edge of the floor of Hermite crater, the southern edges of the floors of Peary and Bosch craters in the northern polar region. "One would have to travel to a distance well beyond the Kuiper Belt to find objects with surfaces this cold," said the Diviner team on their website. "The Moon has one of the most extreme thermal environments of anybody in the Solar System," said Paige.

Robert Heinlein got it right when he dubbed the Moon a harsh mistress.

What Makes Lunar Dust Sticky?

In the 1960s and 1970s, the Apollo Moon program struggled with a minuscule, yet formidable enemy: sticky lunar dust. Four decades plus later, a new study reveals that forces compelling lunar dust to cling to surfaces—ruining scientific experiments and endangering astronauts' health—change during the lunar day with the elevation of the Sun.

The study analyzes the interactions on the Moon among electrostatic adhesive forces, the angle of incidence of the Sun's rays, and lunar gravity. It concludes that the stickiness of lunar dust on a vertical surface changes as the Sun moves higher in the sky, eventually allowing the very weak lunar gravity to pull the dust off.

"Before you can manage the dust, you have to understand what makes it sticky," says Brian O'Brien, the sole author of the paper. His analysis is the first to measure the strength of lunar dust's adhesive forces, how they change during the lunar day—which lasts 710 h—and differ on vertical and horizontal surfaces. O'Brien used data from the matchbox-sized Dust Detector Experiments deployed on the Moon's surface in 1969 during the *Apollo 11* and *Apollo 12* missions.

Lunar dust has long been described as the No. 1 environmental hazard on the Moon. It causes miscellaneous havoc, from destroying scientific equipment deployed on the lunar surface—dusty surfaces absorb more sunlight and make devices overheat—to creating blinding dust clouds that interfere with lunar landings. It also may be a health hazard to space travelers, since dust clinging to spacesuits detaches when astronauts reenter their lunar module. It then floats free in zero gravity, ready to be inhaled, during the 3-day journey back to Earth.

Lunar dust particles are minuscule, with an average size of 70 µm, the thickness of a human hair. The particles get positively charged by photoelectric effects caused by powerful solar ultraviolet radiation and X-rays—the thin lunar atmosphere does not attenuate solar radiation—generating strong electrostatic adhesive forces that compel the specks of dust to cling to surfaces (Fig. 8.15).

In his new study, O'Brien analyzed the behavior of dust on horizontal and vertical solar cells in one of the Apollo dust-detecting

Fig. **8.15** Gene Cernan, in the lunar module, after battling the dusty lunar surface during *Apollo 17*. (Credit: NASA)

experiments. On the first morning of the experiment, the lunar module—130 m (426 ft.) away from the dust detector—took off from the Moon's surface. The blast of exhaust gases completely cleansed a dusty horizontal solar cell, because it was illuminated only by weak early-morning light, and thus the adhesive force of dust was faint. But only half the dust covering the vertical cell was removed by the blast, because its surface faced east—into more intense sunlight—and thus the sticky forces were stronger.

O'Brien found that later, as the Sun rose and the angle of incidence of the Sun's rays on the dusty vertical surface facing east decreased, the electrostatic forces on the vertical cell weakened. The tipping point was reached when the Sun was at an angle of about 45°. At that time the pull of lunar gravity counteracted the adhesive forces and made the dust start falling off. All dust had fallen by lunar night.

"These are the first measurements of the collapse of the cohesive forces that make lunar dust so sticky," O'Brien says.

In 1965, NASA selected O'Brien, an Australian physicist who was then a professor of Space Science at Rice University in Houston, Texas, to be the principal investigator in one of seven lunar experiments designed for the Apollo program. O'Brien started researching lunar dust in 1966 because he feared for an instrument he developed that was to be left behind on the Moon by the *Apollo 14* mission. He worried that the device, which measured the flux of charged particles, would end up covered in dust and ruined. Lunar dust is "a bloody nuisance," he says.

In 1970, O'Brien published a paper that he says proved that rocket exhaust gases from the *Apollo 11* lunar module had lofted dust and debris, which then coated the surface of a lunar seismometer—the first instrument deployed by human hands on a celestial body. The seismometer then overheated by 50° and failed after 3 weeks' operation. An official 1969 NASA report was incorrect in stating that no contamination had occurred, O'Brien says.

However, it wasn't until late 2006, when O'Brien learned from NASA's website that the space agency had misplaced data tapes from its dust-detecting experiments, that he decided to revisit his own set of 173 tapes. NASA had sent him these tapes one by one in 1969 and 1970, when he was working at the Department of Physics at University of Sydney. He took them with him when, in

1971, he moved to Perth for a new job. O'Brien's tapes are now the only known record of data from those vintage experiments.

Working alone and self-funded, the 75-year-old scientist dedicated 2 years to analyzing paper charts printed out in 1969 and 1970 from the magnetic tapes, which contain six million measurements, most of them yet to be analyzed.

For future Moon and Mars missions, O'Brien offers a practical solution to the dust hazard. Use a wide, Sun-proof shed, to block the rays that enhance dust's adhesive forces.

"Getting closer to understanding the physics of the lunar dust problems means moving one giant step towards management of the hazards," O'Brien says.

Artifacts on the Moon?

As nutty as it may seem to the uninitiated, the notion of looking for artifacts on our own Moon may finally be gaining mainstream scientific traction.

There are good reasons to seriously consider the possibility that at some point in the Earth-Moon system's storied 4.5 billion year-old history, an intelligence may have passed through our Solar System, leaving physical artifacts of their visits.

These artifacts would likely entail more than just space trash, and would arguably include evidence of scientific or industrial activity, such as extremely advanced lunar mining, energy generation—even technology related to lunar nearside Earth reconnaissance.

Or so says Paul Davies, a longtime SETI (Search for Extraterrestrial Intelligence) researcher, physicist, and now director of the Beyond Center at Arizona State University in Tempe.

With the success of crowdsourcing, citizen science initiatives such as SETI@home, Einstein@home, and Cosmology@home, however, Davies and a handful of other serious scientific researchers are advocating marrying crowdsourcing analysis with the images now being cataloged by NASA's LRO.

Since 2009, LRO has been measuring lunar landforms down to a half meter resolution, in the process targeting more than 10,000 lunar sites and covering up to 90 % of the lunar surface.

The mission's current success has resulted in a treasure trove of thousands of very high resolution images, almost all of which could be searched via a citizen science initiative.

Davies thinks the ideal lunar survey would not only include a search for optical anomalies but would go beyond the breadth of LRO's own mission to include searches for evidence of lunar industrial activity.

"[Evidence of past] mining or quarrying could show up in gravimetric or magnetic surveys, even if an ancient mine was buried under the lunar regolith," said Davies. "We could detect nuclear waste perhaps from a lunar satellite by looking for localized gamma ray sources from the lunar surface."

A crowd source lunar image analysis initiative might use Tomnod-type search software in the same way that volunteers were recruited to search satellite imaging for the missing Malaysian 777. Davies says at some stage any search needs to be automated and use state-of-the-art software.

"In searching for artifacts, one is looking for 'something fishy,'" said Davies. "But 'fishiness' requires a human decision in advance about a signature of artificiality. There are some simple examples, like right angle edges. But we have little idea what million-year-old technology might look like."

Yet Andrew Siemion, a research astronomer at the University of California at Berkeley, says citizen science projects involving image analysis are relatively straightforward to set up.

"Professional astronomers sometimes suffer from the tendency to discount anything other than our expected signal as instrumental noise or some kind of interference," said Siemion. "When identifying the unexpected, the eye of an amateur citizen scientist can be just as effective, if not more so, than that of a conditioned professional."

Volunteers sifting through images as part of a crowdsourcing effort could make their efforts dual purpose. That is, they might look for orange pyroclastic rocks or even residual volcanism at the same time they would look for artificial anomalies.

Davies says a search for lunar artifacts should be combined with a search for unusual geological features. However improbable, Davies says planetary scientists need to keep their eyes open for non-random anomalies, even ones on the Moon, and scrutinize their respective databases to "keep an eye out" for putative signatures of technology.

A 1995 academic paper by Ukrainian radio astronomer Alexey Arkhipov argues that only artifacts larger than one meter in size would be found on the lunar surface, with objects smaller than that buried by meters of regolith due to the lunar surface's continual bombardment by micrometeorites.

Even so, Davies says, the Moon is an attractive environment to search for artifacts because they would be preserved for much longer than on Earth (or Mars). "On Earth, human artifacts get buried in centuries," said Davies. "On the Moon it takes millions or tens of millions of years."

However, Davies thinks the case for jettisoned material or junk is stronger than a gadget deliberately left for what might be an unknown and truly immense duration.

Arkhipov argued that the peak of the southern wall of the Moon's nearside crater "Malapert" would make a logical site for reconnaissance of Earth, since our planet can always be seen from there.

"Given that the Moon is a big place, it pays to narrow the search by such educated guesses," said Davies. "Lunar lava tubes would preserve artifacts and also provide an attractive location for equipment to be shielded from ultraviolet radiation and meteorites."

When would probes have first arrived in our Solar System? Because Earth is only about a third of the age of the universe, habitable planets in the galaxy could thus have emerged at least 8 billion years ago, says Davies. So, he notes it's likely that if technology ever entered our Solar System, it happened a long time ago. Assuming that the number of technological extraterrestrial civilizations remain uniform over time, then Davies says it's still arguable that our Solar System has been visited at least once during that 8-billion-year timeframe.

Thus, Davies reasons that the average expectation for a visit is of the order 4 billion years ago and later. But he says even a 100 million years ago is optimistically the most recent timeframe for their arrival. And to think they've been here since the dawn of recorded human civilization, he says, would be pretty much a statistical impossibility. However you cut the numbers, says Davies, you would not expect "recent" visits.

In any event, Davies doesn't expect that there have been any visits by flesh and blood entities, and if there were, he reckons they would have moved on. In the event biological entities did travel to actually colonize a new planet, Davies says they would likely pick one without burgeoning life forms, due to difficulties cohabitating with any existing biology.

"My position is that biological intelligence is but a transitory phase in the evolution of intelligence in the universe," said Davies. "Why dispatch fragile biological entities on a hazardous journey across the vastness of space when almost all the intellectual heavy lifting, let alone the physical grunt work, will be done by designed systems?"

And Davies says if such a system happened to enter our Solar System, for reasons we cannot even imagine, it may either "stay, go, or multiply."

Of course, in science fiction, the most famous artifact was the enigmatic monolith envisioned in Arthur C. Clarke's *2001: A Space Odyssey*. In the novel and subsequent film, the monolith appears to reactivate after being found just a few meters under the lunar regolith.

SETI searchers have long considered the possibility that dormant probes may have been sent to our Solar System to wait as silent sentinels before our own technology wakes up to their possible presence. SETI researchers have even considered the possibility of beaming radio beacons to the Earth-Sun gravitational Lagrange points in hopes of "awakening" such unseen probes. That is, if they are out there.

In the last 40 years, before the advent of digital photon counters on telescopes, there were two relatively cursory searches within both the Earth-Sun and Earth-Moon Lagrange points, using comparatively small aperture optical telescopes at Kitt Peak, Arizona, and Leuschner Observatory in California. Both failed to detect any such non-human artificial objects.

John Gertz, president of the California-based FIRSST (Foundation for Investing in Research on SETI Science and Technology) initiative, suggests conducting a radio search for a beacon within our own inner Solar System that would have been activated at the probe's first detection of Earth's own electromagnetic leakage. He thinks it would now be broadcasting now at very low wattage and would have no message, other than the implied one, "I am here; and I am artificial."

Likely smaller than a car, but larger than a grapefruit, Gertz says the payload, a virtual *Encylopedia Galactica*—or their civilization's complete history and knowledge—could be stored on a thumb drive that we would literally have to physically retrieve.

This, of course, assumes that the designers of such probes would have an innate desire to bare their souls to an emerging technology such as ours.

"Physical encoding is a very efficient way of transmitting very large amounts of information from point-to-point," said Siemion. "So, it is entirely possible that an advanced civilization might choose to disseminate large amounts of information via encoded physical artifacts."

There may even be more than one such probe waiting for us. "There is no reason to believe that only one civilization has sent a probe; there may be a variety of probes out there," said Gertz.

So, why not look? "I advocate searching all free searchable databases just for the hell of it," said Davies.

9. Surprising Discoveries

Radioactivity

The upper 8 miles of the Moon's crust are surprisingly radioactive. When *Apollo 15* astronauts used thermal equipment to measure it, they got unusually high readings, which indicated that the heat flow near the Apennine Mountains was rather hot. The core is not hot at all but cold. The amount of radioactive material on the surface is not only "embarrassingly high" but difficult to account for. Where did all this hot radioactive material (uranium, thorium, and potassium) come from?

In a surprising discovery, scientists have found that the Moon itself is a source of potentially deadly radiation. Measurements taken by NASA's LRO show that the number of high-energy particles streaming in from space did not tail off closer to the Moon's surface, as would be expected with the body of the Moon blocking half the sky. Rather, the cosmic rays created a secondary—and potentially more dangerous—shower by blasting particles in the lunar soil, which then became radioactive.

"The Moon is a source of radiation," said Boston University researcher Harlan Spence, the lead scientist for LRO's cosmic ray telescope. "This was a bit unexpected."

Although the Moon blocks galactic cosmic rays to some extent, the hazards posed by the secondary radiation showers counter the shielding effects, Spence told the American Geophysical Union. Overall, future lunar travelers face a radiation dose 30–40 % higher than originally expected, he noted.

Galactic cosmic rays, believed to be caused by supernova explosions, are fast-moving electrically charged electrons and atomic nuclei. They are found throughout the Solar System, though their numbers are particularly high at the present time due to an unusually quiet period of solar activity.

The lack of solar magnetic fields and reduced solar wind pressure means cosmic rays have been able to more easily penetrate

© Springer International Publishing Switzerland 2016
V.S. Foster, *Modern Mysteries of the Moon*, Astronomers' Universe,
DOI 10.1007/978-3-319-22120-5_9

the inner Solar System. "We are in a period when the radiation risks are elevated, but still tolerable," Spence said, adding that the levels were about what an X-ray technician or uranium miner might normally experience in a year.

The Sun cycles through periods of high and low magnetic activity about every 11 years. The most recent period of maximum activity, characterized by increasing number of sunspots and other disruptions of the Sun's surface was in 2013.

LRO found about twice as many of the highest energy cosmic rays—those that could punch through the telescope itself—as expected. NASA plans to use the information to design shelters and operating procedures for future human excursions to the Moon.

LRO's primary mission was to scout for landing spots, find potential resources, and characterize radiation and other hazards. "The work being done in heliophysics areas is important to keeping astronauts safe," noted the late Michael Wargo, who oversaw NASA's lunar research programs before his death in 2013. "One of the holy grails was to be able to predict the Sun's activities and give an "all clear" on how many days astronauts could be on (spacewalks)."

To help accomplish this, the LRO was equipped with the Cosmic Ray Telescope for the Effects of Radiation (CRaTER) detector. Its purpose was to detect radiation in the space around the Moon where the shielding effects of Earth's atmosphere and magnetic field are gone. It also had a scanner that is covered by a special plastic that reacts to radiation in the same way that human muscle tissue does, allowing researchers to observe the effects of deep space radiation on susceptible bone marrow. The data from the instruments was to provide critical information on the radiation hazards that will be faced by astronauts on extended missions to deep space such as those to Mars (Fig. 9.1).

"These data are a fundamental reference for the radiation hazards in near Earth 'geospace' out to Mars and other regions of our Sun's vast heliosphere," says CRaTER principal investigator Nathan Schwadron of the University of New Hampshire Institute for the Study of Earth, Oceans, and Space (EOS).

The space environment poses significant risks to both humans and satellites due to harmful radiation from galactic cosmic rays

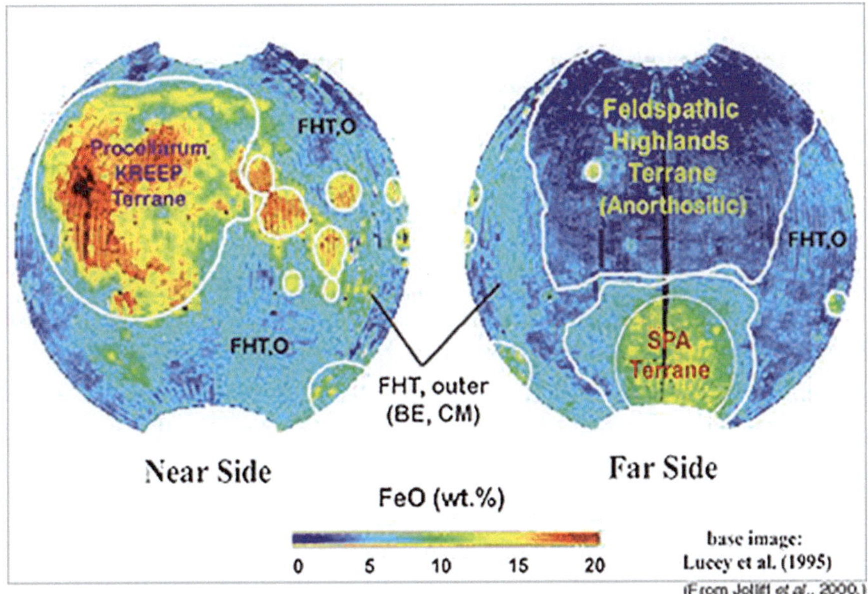

FIG. 9.1 Information on the distribution of radioactivity on the lunar surface was one goal of Lunar Prospector. This map shows that the element thorium is highest on the front side of the Moon, mainly in the highlands south of Mare Imbrium. The correspondence with the Imbrium Basin suggests that the basaltic lavas that filled it were enriched in thorium. Note that corresponding highland surfaces on the far side are lower (Credit: NASA)

and solar energetic particles that can easily penetrate typical shielding and damage electronics. When this radiation impacts biological cells, it can cause cancer.

Before CRaTER's long-term radiation measurements were derived using a material called "tissue-equivalent plastic" (TEP)—a stand-in for human muscle capable of gauging radiation dosage—those hazards were not sufficiently well characterized to determine if long missions outside low-Earth orbit can be accomplished with acceptable risk.

Two pieces of TEP, roughly 2 in. and 1 in. thick, respectively, are separated by silicon radiation detectors. The TEP-detector combo measures how much radiation may actually reach human organs, which may be less than the amount that reaches the spacecraft. Data gathered by the LRO show that lighter materials such as plastics provide effective shielding against the radiation hazards faced by astronauts during extended space travel. The finding could help reduce health risks to humans on future missions into deep space.

Aluminum has always been the primary material in spacecraft construction, but it provides relatively little protection against high-energy cosmic rays and can add so much mass to a spacecraft that it becomes cost-prohibitive to launch.

CRaTER's seminal measurements now provide quantified radiation hazard data from lunar orbit and can be used to calculate radiation dosage from deep space down to airline altitudes. The data will be crucial in developing techniques for shielding against space-based radiation dosage. The measurements have also played a vital role in University of New Hampshire (UNH) space scientists' efforts to develop both the first web-based tool for predicting and forecasting the radiation environment in near-Earth, lunar, and Martian space environments and a space radiation detector that possesses unprecedented performance capabilities.

The near real-time prediction/forecasting tool known as PREDICCS integrates for the first time numerical models of space radiation and a host of real-time measurements being made by satellites currently in space. It provides updates of the radiation environment on an hourly basis and archives the data weekly, monthly, and yearly—an historical record that provides a clear picture of when a safe radiation dose limit is reached for skin or blood-forming organs, for example. CRaTER offers an opportunity to test the capability of PREDICCS to accurately describe the lunar radiation environment.

Equipment also included a detector developed at UNH, known as DoSEN, short for Dose Spectra from Energetic Particles and Neutrons; this measures and calculates the absorbed dose in matter and tissue resulting from the exposure to indirect and direct ionizing radiation, which can change cells at the atomic level and lead to irreparable damage. Schwadron is lead scientist for both the PREDICCS and the DoSEN projects.

Says Harlan Spence, CRaTER deputy lead scientist and director of EOS, "Until now, people have not had the "eyes" necessary to see this particular population of particles. With CRaTER, we just happen to have the right focus to make these discoveries."

Results from CRaTER indicate that while high-energy cosmic rays from deep space known as HZE particles make up only about 1 % of the radiation in the lunar environment, they carry about 50 % of the total energy from radiation.

Radiation in deep space comes from cosmic rays, from the solar wind, and from solar energetic particles emanated during a solar storm. Particles from these sources rocket through space. Many can pass right through matter, such as our bodies. So-called ionizing radiation knocks electrons off of atoms within our bodies, creating highly reactive ions. Within Earth's protective atmosphere and magnetic field, we receive low doses of background radiation every day. The radiation hazards astronauts face are serious yet manageable, thanks to research endeavors such as the CRaTER instrument.

Larry Townsend of the University of Tennessee, who was a co-investigator on the CRaTER team, notes that CRaTER's observations came at a time when solar activity, and hence the solar wind, had been unusually quiet. The solar wind disperses some galactic cosmic rays, but in the current solar lull, more of these rays are able to bombard Earth and the Moon. "They're lower-level exposures" of galactic cosmic rays, Townsend said, "but they're damaging in the sense that the particles are highly charged and heavy, and they create a lot of damage when they're going through the body."

This radiation comes from the partial reflection, also called an albedo, of galactic cosmic rays off the Moon's surface. Galactic cosmic ray protons penetrate as much as a meter (about 3.2 ft) into the lunar surface, bombarding the material within and creating a spray of secondary radiation and a mix of high-energy particles that flies back out into space. This galactic cosmic ray albedo, which may interact differently with various chemical structures, could provide another method to remotely map the minerals present in the Moon's surface.

CRaTER directly measured the proton component of the Moon's radiation albedo for the first time, said Spence. The TEP radiation detector measures various components of radiation separately, which enables CRaTER to, in Spence's words, "unfold" the energy spectrum of the radiation albedo.

This result, he said, illustrates the value of combining exploration and science in spaceflight. "If we had been on a different science-oriented mission, we probably would've developed a different instrument," Spence said. "In fact, we probably never would have flown TEP."

The Lunar Atmosphere

Until recently, most everyone accepted the conventional wisdom that the Moon has virtually no atmosphere. Just as the discovery of water on the Moon transformed our textbook knowledge of Earth's nearest celestial neighbor, recent studies confirm that our Moon does indeed have an atmosphere consisting of some unusual gases, including sodium and potassium, which are not found in the atmospheres of Earth, Mars, or Venus.

It's an infinitesimal amount of air when compared to Earth's atmosphere. At sea level on Earth, we breathe in an atmosphere where each cubic centimeter contains 10,000,000,000,000,000,000 molecules; by comparison the lunar atmosphere has less than 1,000,000 molecules in the same volume. That still sounds like a lot, but it is what we consider to be a very good vacuum on Earth. In fact, the density of the atmosphere at the Moon's surface is comparable to the density of the outermost fringes of Earth's atmosphere where the International Space Station orbits.

What is the Moon's atmosphere made of? We have some clues. The *Apollo 17* mission deployed an instrument called the Lunar Atmospheric Composition Experiment (LACE) on the Moon's surface. It detected small amounts of a number of atoms and molecules including helium, argon, and possibly neon, ammonia, methane, and carbon dioxide (Fig. 9.2).

From here on Earth, researchers using special telescopes that block light from the Moon's surface have been able to make images of the glow from sodium and potassium atoms in the Moon's atmosphere as they are energized by the Sun. Still, we only have a partial list of what makes up the lunar atmosphere. Many other components are expected.

Scientists think that there are several sources for producing gases in the Moon's atmosphere. These include high-energy

FIG. 9.2 Contrary to this humorous ad, the Moon does have an atmosphere

photons and solar wind particles knocking atoms from the lunar surface, chemical reactions between the solar wind and lunar surface material, evaporation of surface material, material released from the impacts of comets and meteoroids, and outgassing.

Outgassing is the release of gases from the lunar interior, usually through radioactive decay. Outgassing events may also occur during moonquakes. After being released, lighter gases escape into space almost immediately. Outgassing replenishes the tenuous atmosphere.

Another source are molecules that are loosened from the surface when other molecules from space hit the ground. These molecules may migrate across the surface of the Moon to colder regions, where they condense again on the ground, or they may fly off into space. This mechanism may be a source of lunar water.

The impact of sunlight, the solar wind, and micrometeorites hitting the Moon's surface can also release gases that were buried in the lunar soil—a process called sputtering. These gases either fly off into space or bounce along the lunar surface. Sputtering may explain how water-ice is collected in lunar craters. Comets hitting the Moon may have left some water molecules on the surface.

Some of the molecules then accumulated in dark polar craters, forming beds of solid ice.

Ultraviolet sunlight affects the released gases by ejecting electrons, which gives them an electrical charge that can cause the particles to levitate more than a mile into the sky. At night, the opposite occurs. Atoms receive electrons from the solar wind and settle back down near the surface. This floating fountain of Moon dust travels along the boundary between night and day (lunar terminator), creating a glow similar to Earth's sunsets. Known as the lunar horizon glow, it was observed several times during Apollo missions.

Molecules from space come from the solar wind. Because its surface is protected by neither an atmosphere nor a magnetosphere, the Moon is constantly exposed to the solar wind, and these molecules get buried in the Moon's surface. It is likely that scientists on Earth will someday understand more about the process of nuclear fusion, which is another way to create energy. Then these molecules buried in the Moon's surface may become an important source of fuel for energy.

With the discovery of significant ice deposits at the Moon's poles by NASA's Lunar CRater Observation and Sensing Satellite (LCROSS) and the Lunar Reconnaissance Orbiter (LRO) missions, and the discovery of a thin scattering of water molecules in the lunar soil by the Chandrayaan X-ray Observatory, another fascinating possibility has captured researchers' interest. The Moon's atmosphere may play a key role in a potential lunar water cycle, facilitating the transport of water molecules between polar and lower latitude areas. The Moon may not only be wetter than we once thought, but also more dynamic.

One of the critical differences between the atmospheres of Earth and the Moon is how atmospheric molecules move. Here in the dense atmosphere at the surface of Earth, the molecules' motion is dominated by collisions between the molecules. However the Moon's atmosphere is so thin, atoms and molecules almost never collide. Instead, they are free to follow arcing paths determined by the energy they receive from the processes described above and by the gravitational pull of the Moon.

The technical name for this type of thin, collision-free atmosphere that extends all the way down to the ground is a "surface boundary exosphere." Scientists believe this may be the most common type of atmosphere in the Solar System. In addition to the Moon, Mercury, the larger asteroids, a number of the Moons of the giant planets, and even some of the distant Kuiper Belt objects out beyond the orbit of Neptune, all may have surface boundary exospheres. But in spite of how common this type of atmosphere is, we know very little about it. Having one right next door on our Moon provides us with an outstanding opportunity to improve our understanding.

To pursue that prospect, NASA launched the Lunar Atmosphere and Dust Environment Explorer (LADEE) (pronounced "laddie")—a robotic mission that orbited the Moon to gather detailed information about the structure and composition of the thin lunar atmosphere, and to determine whether dust is lofted into the lunar sky. A thorough understanding of these characteristics of our nearest celestial neighbor will help researchers understand other bodies in the Solar System, such as large asteroids, Mercury, and the moons of outer planets.

LADEE ran its science instruments almost non-stop at amazingly low orbits, just a few km above the surface. Early results suggest that LADEE was low enough to see some new things, including increased dust density and possibly new atmospheric species.

LADEE's Neutral Mass Spectrometer measured the compositional, temporal, and spatial changes of the lunar exosphere by examining the abundance of a number of species. After argon-40 was discovered in the lunar atmosphere by the Apollo missions, LADEE's NMS instrument confirmed the presence of 40Ar and tracked its variation over time, showing that its abundance varied on a diurnal basis (occurring during the daytime rather than at night). Argon freezes out in the cold temperatures of lunar night and then sublimates at lunar dawn, when the surface is heated by the Sun.

NMS also examined two species that are regularly delivered to the Moon by solar wind—helium and neon-20. Helium is not bound to the Moon and is constantly lost into space and replenished by the solar wind. NMS was able to track and quantify the loss of helium when the Moon was in Earth's geomagnetic tail.

The Lunar Dust Experiment (LDEX) recorded more than 11,000 dust impacts over the primary mission. LDEX has mapped the temporal and spatial distribution of dust events and found that dust events significantly increase at lower altitudes, results that were expected before the mission.

Dust is more abundant over the sunrise sector of the Moon in the direction of motion of the Earth-Moon-System around the Sun. Also, LDEX was able to measure bursts of dust that are likely caused by lunar impacts that release dust into the exosphere that then dissipates over time.

The use of the present lunar "vacuum" for industrial purposes as well as for scientific purposes (e.g., astronomical observations) will most likely necessitate the maintenance of a sufficiently "lunar-like" exosphere rather than allowing a substantial atmospheric mass to build up.

Three sources can be identified within the framework of large-scale lunar operations as potentially releasing substantial quantities of gases into the lunar exosphere: mining and processing of lunar materials, leakages from the Moon base environment, and fuel expenditures during transportation of personnel and materials to and from the lunar surface.

To build the Moon base, raw material is propelled from the lunar surface by a mass launcher and not by the use of chemical rockets. Without materials processing on the lunar surface a potential source of gases is eliminated, and by using nonchemical methods to lift the required lunar materials off the lunar surface, the mass released during transport through the Moon's atmosphere is minimized.

It is believed that mining operations will not release a significant amount of gases into the atmosphere. If it is assumed that 10^{10} kg of lunar materials are mined during a 10-year period and that 10 % of the trapped gases in the lunar soil (10^{-4} to 10^{-5} of its mass) are released during normal mining operations, the average release rate is 3×10^{-4} kg/s, which is substantially less than the natural source rate, according to Richard R. Vondrak, Stanford Research Institute.

Losses due to leakage from a Moon base have been estimated by NASA experts based upon a projected loss rate per unit surface area. The yearly leakage loss is approximated to be 18,000 kg, which would result in a release rate of 6×10^{-4} kg/s. Again this is insignificant in comparison to the natural source rate. It should also be noted that the Moon base considered by Nishioka et al. includes a processing plant and would most likely be larger than the lunar facility considered here. The actual release rate would then be even smaller than that given above.

By far the most significant source for release of gases into the lunar environment are the exhaust products released by chemical rockets in the initial establishment of the lunar base and its continual resupply. It has been estimated that 1 kg of propellant will be expended by the lunar landing vehicle for each kg of payload landed. The mass of a lunar base, estimated at 17×10^6 kg, is assumed to be delivered to the lunar surface over a 2.5-year construction phase. An annual resupply rate of 0.4–0.5×10^6 kg has been calculated. These figures give release rates of 0.2 kg/s during the establishment of the base and 0.02 kg/s thereafter by averaging the expenditures of propellant over an entire year. This is believed to be valid due to the rapid diffusion of gases released on the lunar surface, according to Vondrak.

Although establishing a lunar base will most likely result in gas release rates at many times greater than that occurring naturally, a sparse lunar exosphere will still be preserved, given the magnitudes of the calculated release rates. Furthermore, a long-lived atmosphere will not result, and if the critical sources of gas are halted, the lunar atmosphere will return to its natural state within weeks.

Vanishing Magnetic Field

Early lunar tests and studies indicated that the Moon had little or no magnetic field. Then lunar rocks returned to Earth proved to be strongly magnetized. This was shocking to scientists, who had not expected these very strange magnetic properties. NASA cannot explain where this magnetic field came from.

The Moon has an external magnetic field that is very weak in comparison to that of Earth. Other major differences are that the Moon does not currently have a dipolar magnetic field (magnetic field between poles), and the varying magnetization that is present is almost entirely crustal in origin.

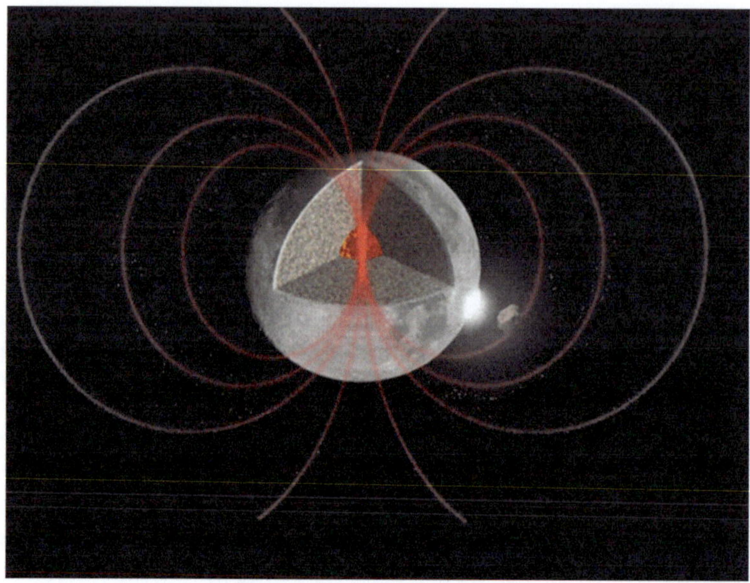

Fig. 9.3 This illustration shows one suggested mechanism for creating an ancient magnetic field on the Moon. In this scenario, impacting space rocks on the Moon would create instability in the Moon's core that could lead to a dynamo that creates a magnetic field

One hypothesis holds that the crustal magnetizations were acquired early in lunar history when a geodynamo (process through which a rotating, convecting, and electrically conducting fluid can maintain a magnetic field over astronomical time scales) was still operating. The small size of the lunar core, however, is a potential obstacle to promoting that hypothesis (Fig. 9.3).

Alternatively, it is possible that on an airless body such as the Moon, transient magnetic fields could be generated during large impact events. In support of this, it has been noted that the largest crustal magnetizations appear to be located near the antipodes of the giant impact basins. It has been proposed that such a phenomenon could result from the free expansion of an impact-generated plasma cloud around the Moon in the presence of an ambient magnetic field.

For example, the Chandrayaan-1 spacecraft mapped a "mini-magnetosphere" at the Crisium antipode on the Moon's far side, using its Sub-keV Atom Reflecting Analyzer (SARA)

instrument. The mini-magnetosphere is 360 km across at the surface and is surrounded by a 300-km-thick region of enhanced plasma flux that results from the solar wind flowing around the mini-magnetosphere.

There's growing evidence that fine particles of moondust might actually float, ejected from the lunar surface by electrostatic repulsion. This could create a temporary nighttime "atmosphere" of dust. The moondust atmosphere might also gather itself into a sort of diaphanous wind. Drawn by differences in global charge accumulation, floating dust would naturally fly from the strongly negative night side to the weakly negative day side.

This "dust storm" effect would be strongest at the Moon's terminator. Much of these details are still speculative, but the Lunar Prospector spacecraft detected changes in the lunar night side voltage during magnetotail crossings, jumping from –200 to –1000 V.

The plasma sheet is a very dynamic structure, in a constant state of motion, so as the Moon orbits through the magnetotail the plasma sheet can sweep across it many times, with encounters lasting anywhere from minutes to hours or even days.

The Moon generated a surprisingly intense magnetic field until at least 3.56 billion years ago, 160 million years longer than previously thought, a new study reports. These findings could shed light not just on the magnetic field of the Moon, which is now extremely weak, but on that of asteroids and other distant worlds, investigators added.

Earth's magnetic field is created by its internal dynamo, which itself is generated by the planet's churning molten metal core. Research increasingly suggests that the Moon once had a dynamo as well, with evidence of magnetism found in lunar rocks returned by Apollo astronauts.

A magnetic field is generated by a dynamo, which is caused by the fluid motion of a conducting material, such as liquid iron. In the case of Earth's magnetic field, this motion occurs in the planet's outer core, and is caused by the convection of heat. But the Moon isn't large enough for convection to take place. Scientists were at a loss to explain what else might generate the required liquid motion of iron inside the Moon—until now.

In one new proposal, Christina Dwyer of the University of California, Santa Cruz, and her colleagues suggest that the Moon's solid-rock middle layer, its mantle, stirs up its liquid iron core. The researchers think this happens because the Moon's core and its mantle rotate around slightly different axes, and the boundary between them is not quite spherical, so their relative motion causes the fluid to mix around.

The strength of this stirring is determined by the angle between the core and the mantle, and the distance between Earth and the Moon, because the tidal gravitational tug from Earth causes the Moon's mantle to rotate differently than the core.

This model would explain why the Moon used to have a magnetic field but no longer does. It's because the angle between the mantle and the core has narrowed over time, while the distance between the Moon and Earth has widened, causing the tidal forces to steadily decrease. Although these forces used to be enough to generate a dynamo inside the Moon, they aren't anymore. Based on their calculations, the researchers estimate the lunar magnetic field might have lasted for about a billion years, somewhere between around 2.7 billion and 4.2 billion years ago.

"Based on what we know about stirring, and everything we know about fluid motion, we can find no reason that this would not work," Dwyer said. "All the flags are go, and now this needs to be taken to the next level to get tested."

Dwyer said her research team had studied the basic scenario but hopes that scientists who specialize in modeling the complex physics of dynamos will now step in to investigate whether this could explain what happened on the Moon. But theirs isn't the only possible solution to this Moon mystery.

Michael Le Bars of the French National Center for Scientific Research and the Université Aix-Marseille in France, along with his colleagues, propose another explanation for the ancient lunar magnetic field.

Le Bars' team also suggests that the Moon's mantle might have stirred up the liquid in its core. However, it offers a different impetus for this stirring. Instead of tidal interactions between Earth and the Moon, the researchers posit that impacts by large space rocks slamming into the Moon have changed its rotation rate, causing differential motion between the mantle and the core.

Although the first scenario would cause steady stirring while the Earth and Moon were at the right distance apart, the second picture would induce brief periods of especially strong stirring of the core, creating spikes of a magnetic field on the Moon.

Either option could be correct, but it's also possible that both mechanisms played a role in causing an ancient magnetic field on the Moon.

"The two studies are thought-provoking and may be complementary," Dominique Jault, a researcher at ETH Zürich in Switzerland and the Université Joseph-Fourier in France, who was not involved in either new study, wrote in an accompanying essay in the same issue of *Nature*. "Future paleomagnetic experiments on samples from very old lunar rocks will enable their theories to be tested."

A new analysis of the biggest lunar rock brought back to Earth by *Apollo 11* astronauts in 1969 reveals the Moon's dynamo lasted about 160 million years longer than previously thought, well after the last of the largest crater-forming impacts hit the Moon. Scientists investigated a 5-g (0.18 of an ounce) sample taken from a 3.56-billion-year-old volcanic Moon rock from the Sea of Tranquility.

"When rocks solidify from lava, they capture a record of the magnetic field in their environment," said study lead author Clément Suavet, a geoscientist at MIT. "By studying rocks of different ages, we can reconstruct the history of lunar-surface magnetic fields."

The analysis revealed that the intensity of the lunar magnetic field was exceptionally strong 3.56 billion years ago, "almost identical to the field measured in a previous study of 3.7-billion-year-old rocks," Suavet said. "This seems to indicate that the lunar magnetic field was remarkably stable."

The ancient magnetic field of the Moon was about as intense as Earth's current surface magnetic field. This makes it about 1000 times stronger than the Moon's present surface magnetic field, researchers said.

Learning more about how the Moon's dynamo originated and developed could yield insights into the dynamos of smaller objects, such as asteroids, and larger bodies, such as planets.

"The Moon is like a giant laboratory where we can test our theories about how planets form and evolve," Suavet said. Many questions remain about the Moon's magnetic field, such as why it was so intense late into lunar history and how and why it disappeared over time. "The question is, when and how did the dynamo decay?" Suavet said.

In a separate study, scientists have recently discovered another major factor contributing to the magnetic field of the Moon. When the Moon is full, it develops a strong electrical field near the surface as it swings through Earth's magnetic "tail," according to new observations from a Japanese probe.

Earth's magnetic field creates a protective bubble known as the magnetosphere, which surrounds the planet and shields us from solar wind—a rush of charged particles, or plasma, constantly streaming from the Sun.

As the solar wind pushes on Earth's magnetic bubble, the planet's magnetosphere stretches, forming what's called the magnetotail. This tail reaches beyond the orbit of the Moon, and it's always pointed away from the Sun. Meanwhile, we see a full Moon when the lunar orb is on the opposite side of Earth from the Sun—and therefore within the magnetotail.

The strong electric field near the surface was discovered by the Kayuga (Selene) lunar orbiter, managed by the Japan Aerospace Exploration Agency (JAXA).

JAXA launched the Kaguya probe in 2007. The craft orbited at a mere 62 miles (100 km) above the Moon's surface for 20 months, returning the first high-definition movies of the lunar landscape.

Scientists are now mining the data sent back by the craft's imagers and instruments, including the Magnetic field and Plasma experiment (MAP), which recorded that relatively high-energy electrons gyrating in the magnetic field are being absorbed by the lunar surface when the Moon is full.

The result is a strong electric field that develops around the lunar surface the same time as the full Moon. Based on Kaguya's data, the team says this relatively intense electrical field can be found when the Moon passes through the region of the magnetosphere called the plasma sheet, which runs down the middle of Earth's magnetotail.

Since the Moon has no global magnetic field of its own, its surface remains exposed to the trapped charged particles inside the plasma sheet. "We are suggesting the presence of a relatively strong electric field around the Moon when it is inside Earth's magnetosphere, but the origin of this electric field remains a mystery," said study leader Yuki Harada, a graduate student at Kyoto University in Japan.

"We think it may be related to the properties of the plasma and magnetic field in Earth's magnetotail, and also the interaction between the Moon and surrounding plasmas." The newly seen effect adds to what scientists know about the Moon's electrical activity. For example, previous observations have shown that the Moon gets slightly electrified by sunlight, regardless of whether the Moon is in Earth's magnetotail.

The day side of the Moon becomes positively charged, as solar radiation knocks electrons from the surface. Meanwhile, electrons build up on the night side of the Moon and give the surface a negative charge, Harada said.

The big question now is whether a strongly charged lunar surface poses risks to future robotic and human explorers. "It is quite possible that electric fields induce a charge-up and subsequent discharge around a space vehicle, which could bring about serious damages to the human missions," Harada said.

The biggest hazards of an electrified Moon may be that the static charge can transport large amounts of the Moon's abrasive dust, which can damage sensitive lenses and electronics. The static buildup can also lead to unexpected electrical discharges.

NASA's Explorer series of probes and the Apollo missions were the first to reveal the perplexing lunar plasma environment, hinting that electrically driven dust may be a concern to robots and humans setting down on the Moon.

"We've been to the surface before and survived just fine, but we did have a number of problems with dust, among other things. And we happened to be there during very quiet plasma conditions," said Jasper Halekas, a plasma physicist at the University of California, Berkeley.

"Things might be very different during a solar storm, or during a passage through the plasma sheet, the region that was looked at in this study," he said. "Certainly when you have big electric fields, you start to worry about damage to sensitive electronics, etc. And if those electric fields mobilize dust, that could become an additional problem."

But, he said, "the truth is that we don't really know yet what relevance these kinds of studies may have for exploration. Probably the only way we will ever know for sure is to go back to the surface."

For the first time, scientists have observed a mini magnetosphere on the Moon—a magnetic field "bubble" that protects part of the lunar surface from punishing solar radiation.

This mini magnetosphere lies near the Moon's equator, where sunlight reaches only half the time. This would limit the power available to charge solar arrays, should humans ever return to the Moon. But there's evidence other magnetospheres might lie in more favorable landing sites, offering some radiation protection for astronauts.

Earth is fully enveloped in its magnetosphere, because our planet's solid iron inner core rotates inside a molten outer core, creating a magnetic dynamo that generates a global magnetic field.

Extending up to 52,000 miles (78,000 km) above the surface, Earth's magnetic field traps and redirects the constant stream of charged particles from the Sun, aka the solar wind, which could otherwise fry life as we know it.

Sometimes, when a burst of solar activity overloads Earth's magnetosphere, charged particles seep through and interact with our atmosphere. "I live far north and can look up and see the [resulting] aurorae," said study co-author Martin Wieser, a senior scientist at the Swedish Institute of Space Physics. By contrast, the Moon doesn't have a global magnetic field, so its surface is constantly being bombarded by the Sun's charged particles. However, new data from India's Chandrayaan-1 lunar probe have confirmed that the Moon does have a miniature version of a magnetosphere covering a small pocket in the northeastern region of the side of the Moon that faces away from Earth.

When Wieser and his team measured it, the Moon's mini magnetosphere was 224 miles (360 km) across. The magnetosphere acts like a stone in a stream, causing the solar wind to move faster and get thicker in a 186-miles (300-km) ring around the obstacle. The magnetic field at the center of the bubble is 300 times weaker than the section of Earth's magnetic field over the equator and 600 times weaker than the field over Earth's poles.

The origins of the mini magnetosphere are puzzling, although some scientists have noted that magnetic fields tend to appear on the opposite side of the Moon from large impact craters.

"There is a theory that links the two facts. Plasma [charged gas] created on a large impact would flow around the Moon and freeze in the magnetic field when the plasma wave meets on the other side of the Moon," Wieser said.

"But there are also magnetic anomalies that do not match this pattern that are more difficult to explain." Previous observations had hinted that mini magnetospheres might exist on the Moon, but the Chandrayaan-1 data showed the effect much more clearly. That's because the probe was able to "see" the behavior of particles much closer to the Moon's surface.

Similar mini magnetospheres elsewhere on the Moon could explain mysterious bright "swirls" on the lunar surface, Wieser said. Solar radiation hitting the lunar soil darkens the surface over time, but magnetospheres might be protecting such regions from this dimming effect.

The discovery of the Moon's first mini magnetosphere opens the door for finding magnetospheres on smaller bodies, even asteroids, Wieser said.

Most asteroids would be dwarfed by even a mini magnetosphere, Wieser said, so "they will be completely inside such a magnetospheric bubble. If the field is strong enough, this could alter significantly how the asteroid evolves."

Understanding asteroid evolution can in turn tell us more about the birth of the Solar System, he pointed out. Asteroids lack atmospheres and therefore are not subject to most of the erosive processes that alter evidence of geologic histories on other bodies.

"These objects are thus thought to be much closer to how the Solar System was in the beginning," he said. Knowing the effects of solar wind on asteroids and how magnetic anomalies might play a role will be an important link in the story.

"Any process modifying the surface over geological times may significantly change our understanding of these objects ... and may therefore change our understanding of the formation and evolution of the Solar System," Wieser said.

Sleeplessness

Next time you can't get to sleep, take a peek outside. If there's a full Moon, it might not be your partner's snoring keeping you awake—it could be your inner caveman.

Scientists claim that we have an internal biological clock hardwired into our genes, and we are geared to sleep less when the Moon is full. We take longer to nod off, and when we finally do our sleep is lighter. But it has nothing to do with the Moon's glow, or even its gravitational pull. Instead, researchers believe the phenomenon is down to the 'circalunar clock' that ticks away according to the Moon's cycle. They suggest lighter sleep would have helped protect our ancestors from predators out hunting during a full Moon.

Dr. Silvia Frey of the University of Basel said: "Our findings are the first that point to the existence of a circalunar clock in humans, but nobody has pinpointed it yet in the body. We expect it works at a molecular level and in the brain, possibly in the hypothalamus, the same part that regulates the circadian rhythm—the body's own internal clock. Originally, centuries ago, it would have made sense if you didn't sleep as much during a full Moon, when there was a lot of light and a higher risk of being predated. In our ancient ancestors, it suggests this behavior would have been a protective feature."

These were the findings of a team of investigators, including Frey, from the University of Basel, the Swiss Federal Institute of Technology and the Switzerland Centre for Sleep Medicine, who recruited 33 volunteers and studied them in a sleep lab on and off over the course of 3 years starting in 2000. The investigators gathered a range of data—brain wave activity during sleep as measured by electroencephalograms (EEG); levels of melatonin, a sleep-related hormone; the amount of time it took subjects to fall asleep and the amount of time they spent in deep sleep; and their subjective reports of how rested they felt the next day. All of it was intended to learn more about human sleep patterns in a general way and, more specifically, how they are affected by age and gender. Only a decade later did the investigators realize that they may be able to re-crunch the data to learn about the Moon.

Thus should all great science be done since, as it turned out, the second look revealed intriguing patterns. On average, the subjects in the study took 5 min longer to fall asleep on the three or four nights surrounding a full Moon, and they slept for 20 fewer minutes. In addition, EEG activity related to deep sleep fell 30 %, melatonin levels were lower, and the subjects reported feeling less refreshed the next day than on other days. The subjects slept in a completely darkened lab with no sight of the Moon, and none of them—at least from what was known—appeared to have given any thought at all to lunar cycles. And since the Moon was not an experimental variable in the original study, it was never mentioned either to the subjects or even among the investigators. "The aim of exploring the influence of different lunar phases on sleep regulation was never *a priori* hypothesized," they wrote in a wonderfully candid passage in their paper. "We just thought of it after a drink in a local bar one evening at full Moon."

In terms of scientific reliability, all of this is both good and not so good. A study can't get more effectively double-blind than if no one is even thinking about the thing you wind up testing for, which makes the findings uniquely objective. On the other hand, the ideal Moon study would have been carefully set up to give equal weight to every night in the lunar cycle. This study—while capturing most of the nights in the month—did so in a less rigorous way.

"The *a posteriori* analysis is a strength and a weakness," concedes lead author Christian Cajochen, head of the University of Basel's Centre for Chronobiology. "The strength is that investigators and subject expectations are not likely to influence the results, yet the weakness is that each subject was not studied across all lunar phases."

Even if the Moon has as significant an effect on sleep as the study suggests, what's less clear is the mechanism behind it. Dark labs eliminate the variable of light, so that can't be it. And before you ask, no, it's not gravity either. The authors stress that while lunar gravity does indeed raise tides in the oceans, it doesn't on lakes and even many seas. Those bodies are simply too small to feel the effects—to say nothing of human bodies.

Rather, the answer is simply that we, like every other species on Earth, evolved on a particular planet with a particular set of astronomical cycles—day and night, full Moons and less full—and our circadian systems adapted. It's hard to say where the internal clock is in, say, a flowering plant, but in humans, it's likely in the suprachiasmatic nuclei, a tiny region of the brain near the optic nerve involved in the production of melatonin, certain neurotransmitters, and other time-keeping chemicals, all in a rhythm consistent with both its terrestrial and cosmic surroundings. Physically, human beings may be creatures of just this world, but our brains—and our behavior—appear to belong to two.

This internal clock is thought to exist in many animals, particularly marine species. Evidence of the phenomenon has been found in marine midges and iguanas on the Galapagos Islands. It has long been thought that the full Moon has an effect on humans, from influencing fertility to our mental state. In fact, the term "lunatic" derives from *luna*, the Latin word for Moon.

Scientists believe the internal clock may work in conjunction with the light-directed circadian clock, which regulates body functions. People have long reported that it is harder to get to sleep and remain asleep when the Moon is full, and even after a seemingly good night's rest, there can be a faint sluggishness—a sort of full-Moon hangover—that is not present on other days. If you're sleeping on the prairie or in a settler's cabin with no shades, the simple presence of moonlight is an inescapable explanation. But long after humans moved indoors into fully curtained and climate-controlled homes, the phenomenon has remained.

What's never been clear is whether it's the real deal—if the Moon really does mess with us—or if it's some combination of imagination and selective reporting, with people who believe in lunar cycles seeing patterns where none exist. Now, this new research suggests that the believers have been right all along.

The human menstrual cycle is the best-known example of the way our bodies—over millions of years of evolution—have synchronized themselves to the rhythms of the Moon. Less well-known is the lunar link to the electrochemistry of the brain in epileptic patients, which changes in the few days surrounding a new Moon, making seizures more likely. And then there are the anecdotal accounts of the effects the Moon has on sleep.

Numerous studies through the years have attempted to prove or disprove the hypothesis that lunar phases affect human sleep.

But results have been hard to repeat. A group of researchers at the famed Max Planck Institute in Germany and elsewhere analyzed data from more than 1000 people and 26,000 nights of sleep, only to find no correlation.

International researchers are being urged to publish their results in hopes of getting to the bottom of the question. Michael Smith and his co-researchers at Sahlgrenska Academy in Sweden have analyzed data generated by a previous sleep study and compared them with the lunar cycle. Based on a study of 47 healthy 18–30 year-olds and published in *Current Biology*, the results support the theory that a correlation exists.

"Our study generated findings similar to the Swiss project," Smith says. "Subjects slept an average of 20 min less and had more trouble falling asleep during the full Moon phase. However, the greatest impact on REM sleep appeared to be during the new Moon."

The retrospective study by the Gothenburg researchers suggests that the brain is more susceptible to external disturbances when the Moon is full. "The purpose of our original study was to examine the way that noise disturbs sleep," Smith continues. "Re-analysis of our data showed that sensitivity, measured as reactivity of the cerebral cortex, is greatest during the full Moon."

Greater cortical reactivity was found in both women and men, whereas only men had more trouble falling asleep and slept less when the Moon was full. Skeptics warn that both age and gender differences may be a source of error, not to mention more subtle factors such as physical condition and exposure to light during the day.

Though fully aware of the issues, Smith is not prepared to dismiss the results of the Gothenburg study. "The rooms in our sleep laboratories do not have any windows," he says. "So the effect we found cannot be attributable to increased nocturnal light during full Moon."

Thus, there may be a built-in biological clock that is affected by the Moon, similar to the one that regulates the circadian rhythm. But all this is mere speculation. Additionally, more highly controlled studies that target these mechanisms are needed before more definitive conclusions can be drawn.

New Rock Stars

Many of us think of our nearest celestial neighbor as having few secrets to give up. However, we are still finding out new things about the Moon. The latest discovery is of two types of Moon rock not seen before. The last time scientists identified a different type of Moon rock was in the 1970s. One of the new types of rock was found on the far side of the Moon, but the other was found on the near side.

One of the most interesting things about the rock found on the far side of the Moon is that it contains magnesium spinel, which has not been seen on the Moon yet. On Earth, a large amount of this spinel is considered a gemstone. The new Moon rock is concentrated in an area that also lacks minerals that scientists have come to expect on the Moon. The discovery of this strange rock might mean a change in how we look at the Moon, according to Dr. Carle Pieters, the scientist who identified the rock.

So this is a big mystery, and it is a very exciting one because now we have to reexamine our understanding of the character of the lunar crust, in particular to the depths that might have been tapped by this enormous basin and that we are now looking at as exposed on the surface.

These are old surfaces that have been undisturbed but have an extremely unusual composition. And even the space weathering that has occurred on the surface throughout the billions of years of history on the Moon has not erased their unusual compositions. So, they are unusual for the kind of compositions we see, but they are also unusual because they have no defining property that allows us to identify them in our imagery, which is quite unusual for features on the surface of the Moon.

It may not be so surprising to some that a new Moon rock discovery was made on the far side of the Moon. But the second Moon rock was found on the near side of Moon. The scientist in charge of identifying the Moon rock on the near side of the Moon, Dr. Jessica Sunshine, knew about the discovery made by Pieters. However, when Sunshine looked at a rock sample from an area with "dark mantle" (from fire fountain deposits during the Moon's earlier geologic activity), she discovered that it contained chrome.

It came from explosive eruptions of lava and gas over large areas of the Moon, about the size of Massachusetts. And we knew that three of them were there; it just turned out that one of them was compositionally different from the other ones, and in particular it had the kind of spinel which is a chromite, because it has chrome in it, and now we're busy trying to figure out why this deposit is different from the one next door, and what it means.

Clearly, whatever we think we know about the Moon, Moon rocks, and the formation of the Moon, there are still mysteries to solve.

Our Incredible Shrinking Moon

Newly discovered cliffs in the lunar crust indicate that the Moon shrank globally in the geologically recent past and might still be shrinking today, according to a team analyzing new images from NASA's LRO spacecraft. The results provide important clues to the Moon's recent geologic and tectonic evolution.

The Moon formed in a chaotic environment of intense bombardment by asteroids and meteors. These collisions, along with the decay of radioactive elements, made the Moon hot. The Moon cooled off as it aged, and scientists have long thought the Moon shrank over time as it cooled, especially in its early history. The new research reveals relatively recent tectonic activity connected to the long-lived cooling and associated contraction of the lunar interior (Fig. 9.4).

"We estimate these cliffs, called lobate scarps, formed less than a billion years ago, and they could be as young as a hundred million years," said Dr. Thomas Watters of the Center for Earth and Planetary Studies at the Smithsonian's National Air and Space Museum in Washington, D. C. Although ancient in human terms, it is less than 25 % of the Moon's current age of more than 4 billion years. "Based on the size of the scarps, we estimate the distance between the Moon's center and its surface shrank by about 300 ft," said Watters, lead author of a paper on this research appearing in *Science*.

"These exciting results highlight the importance of global observations for understanding global processes," said Dr. John

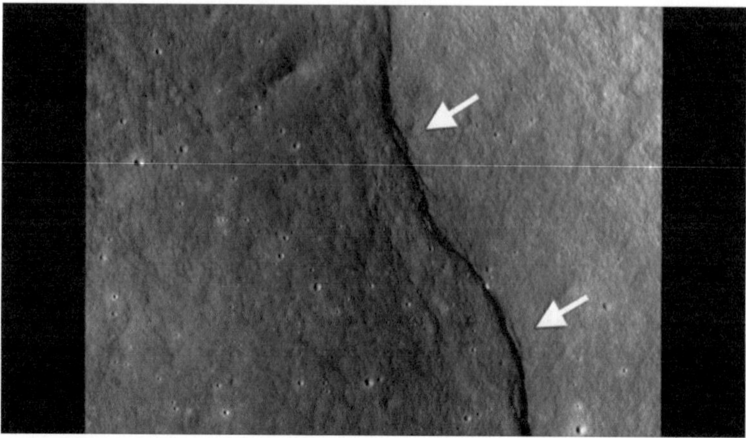

Fig. 9.4 Newly discovered cliffs in the lunar crust indicate the Moon shrank globally in the geologically recent past and might still be shrinking today, according to a team analyzing new images from NASA's LRO spacecraft. The results provide important clues to the Moon's recent geologic and tectonic evolution (Credit: NASA)

Keller, Deputy Project Scientist for LRO at NASA's Goddard Space Flight Center, Greenbelt, MD. "As the LRO mission continues into a new phase, with emphasis on science measurements, our ability to create inventories of lunar geologic features will be a powerful tool for understanding the history of the Moon and the Solar System."

The scarps are relatively small; the largest is about 300 ft high and extends for several miles or so, but typical lengths are shorter, and heights are more in the tens of yards range. The team believes they are among the freshest features on the Moon, in part because they cut across small craters. Since the Moon is constantly bombarded by meteors, features such as small craters (those less than about 1200 ft across) are likely to be young because they are quickly destroyed by other impacts and don't last long. So, if a small crater has been disrupted by a scarp, the scarp formed after the crater and is even younger. Even more compelling evidence is that large craters, which are likely to be old, don't appear on top any of the scarps, and the scarps look crisp and relatively under-graded.

Lobate scarps on the Moon were discovered during the Apollo missions with analysis of pictures from the high-resolution panoramic camera installed on *Apollo 15, 16,* and *17*. However, these missions orbited over regions near the lunar equator, and were only able to photograph some 20 % of the lunar surface, so researchers couldn't be sure the scarps were not just the result of local activity around the equator. The team found 14 previously undetected scarps in the LRO images, seven of which are at high latitudes (more than 60°). This confirms that the scarps are a global phenomenon, making a shrinking Moon the most likely explanation for their wide distribution, according to the team.

As the Moon contracted, the mantle and surface crust were forced to respond, forming thrust faults where a section of the crust cracks and juts out over another. Many of the resulting cliffs, or scarps, have a semi-circular or lobe-shaped appearance, giving rise to the term "lobate scarps." Scientists aren't sure why they look this way; perhaps it's the way the lunar soil (regolith) expresses thrust faults, according to Watters.

Lobate scarps are found on other worlds in our Solar System, including Mercury, where they are much larger. "Lobate scarps on Mercury can be over a mile high and run for hundreds of miles," said Watters. Massive scarps like these lead scientists to believe that Mercury was completely molten as it formed. If so, Mercury would be expected to shrink more as it cooled, and thus form larger scarps, than a world that may have been only partially molten with a relatively small core. Our Moon has more than a third of the volume of Mercury, but since the Moon's scarps are typically much smaller, the team believes the Moon shrank less.

Because the scarps are so young, the Moon could have been cooling and shrinking very recently, according to the team. Seismometers put in place by the Apollo missions have recorded moonquakes. Although most of these can be attributed to things such as meteorite strikes, Earth's gravitational tides, and day/ night temperature changes, it's remotely possible that some moonquakes might be associated with ongoing scarp formation, according to Watters. The team plans to compare photographs of scarps by the Apollo panoramic cameras to new images from the LRO to see if any have changed over the decades, possibly indicating recent activity.

Although Earth's tides are most likely not strong enough to create the scarps, they could contribute to their appearance, perhaps influencing their orientation, according to Watters. During the next few years, the team hopes to use the LRO's high-resolution Narrow Angle Cameras (NACs) to build up a global, highly detailed map of the Moon. This could identify additional scarps and allow the team to see if some have a preferred orientation or other features that might be associated with Earth's gravitational pull.

"The ultrahigh resolution images from the NACs are changing our view of the Moon," said Dr. Mark Robinson of the School of Earth and Space Exploration at Arizona State University, Tempe, AZ, a coauthor and Principal Investigator of the Lunar Reconnaissance Orbiter Camera. "We've not only detected many previously unknown lunar scarps; we're also seeing much greater detail on the scarps identified in the Apollo photographs."

How Fast Does Moon Dust Pile Up?

When Neil Armstrong took humanity's first otherworldly steps in 1969, he didn't know what a nuisance the lunar soil beneath his feet would prove to be. The scratchy dust clung to everything it touched, causing scientific instruments to overheat and, for *Apollo 17* astronaut Harrison Schmitt, a sort of lunar dust hay fever. The annoying particles even prompted a scientific experiment to figure out how fast they collect, but NASA's data got lost.

Or, so NASA thought. Now, more than 40 years later, scientists have used the rediscovered data to make the first determination of how fast lunar dust accumulates. It builds up unbelievably slowly by the standards of any Earth-bound housekeeper, their calculations show—just fast enough to form a layer about a millimeter (0.04 in.) thick every 1000 years. Yet, that rate is ten times previous estimates. It's also more than speedy enough to pose a serious problem for the solar cells that serve as critical power sources for space exploration missions.

"You wouldn't see it; it's very thin indeed," said University of Western Australia Professor Brian O'Brien, a physicist who developed the experiment while working on the Apollo missions in the 1960s and now has led the new analysis. "But, as the Apollo astronauts learned, you can have a devil of a time overcoming even a small amount of dust." That faster-than-expected pile-up also implies that lunar dust could have more ways to move around than previously thought, O'Brien added.

In his experiment, dust collected on small solar cells attached to a matchbox-sized case over the course of 6 years, throughout three Apollo missions. As the granules blocked light from coming in, the voltage the solar cells produced dropped. The electrical measurements indicated that each year 100 µg of lunar dust collected per square centimeter. At that rate, a basketball court on the Moon would collect roughly 450 g (1 lb) of lunar dust annually.

Comparing the effects on cells from dust and from damaging high-energy radiation from the Sun, O'Brien found that long-term dust accretion could diminish the output from shielded power supplies of a lunar outpost more than even the most intense solar outbursts.

Because the threat posed by radiation damage was recognized early on, solar-cell makers fortified their devices against that sort of harm. Yet, "while solar cells have become hardier to radiation, nothing really has been done to make them more resistant to dust," said O'Brien's colleague on the project Monique Hollick, who is also a researcher at the University of Western Australia, was working on in Crawley. "That's going to be a problem for future lunar missions." The work is detailed in *Space Weather*, a publication of the American Geophysical Union.

Before *Apollo 11* blasted off to the Moon in 1969, NASA scientists realized the lunar module would likely kick up a large amount of lunar soil on takeoff, potentially coating nearby science experiments with dust. Detachable covers would require either a small explosive or a physical mechanism to remove after the astronauts left, creating more engineering challenges and room for failure.

"Then I asked what I thought was a pretty common sense question," recalled O'Brien. "If we've got to guard ourselves against damage from the lunar module taking off, who's measuring whether any damage actually took place; who's measuring the dust?"

O'Brien proceeded to quickly invent the Lunar Dust Detector experiment as a small add-on device to the larger experiments. Requiring little power and weighing only 270 g (0.6 lb), the dust detector reported back to Earth alongside the non-scientific housekeeping data. "It really got a free ride," O'Brien said.

The detectors flown on *Apollo 12, 14* and *15* operated until NASA shut them off in September 1977 due to budgetary concerns. Although the detectors worked properly, NASA did not preserve the archival tapes of the data they collected. For three decades NASA assumed the dust detector data had been lost forever, until 2006 when O'Brien heard about NASA's mistake and told them he still had a set of backup copies.

The detectors in the experiment each had three solar cells, all covered with a different amount of shielding against incoming radiation. By comparing damage to the unshielded and shielded solar cells, O'Brien made his determination that dust, rather than radiation, caused the most degradation to the protected cells.

Previous model-based estimates of lunar dust accumulation assumed the dust came entirely from meteor impacts and falling cosmic dust. "But that's not enough to account for what we measured," O'Brien said.

With no atmosphere for wind, the Moon's soil should be stagnant. However, O'Brien said a popular idea of a "dust atmosphere" on the Moon could explain the difference. The concept goes that, during each lunar day, solar radiation is strong enough to knock a few electrons out of atoms in dust particles, building up a slight positive charge. On the nighttime side of the Moon, electrons from the solar wind strike dust particles and give them a small negative charge. Where the illuminated and dark regions of the Moon meet, electrical forces could levitate this charged dust, potentially lofting grains high into the lunar sky.

"Something similar was reported by Apollo astronauts orbiting the Moon who looked out and saw dust glowing on the horizon," said Hollick (Fig. 9.5).

The idea of levitating lunar dust could soon be confirmed by NASA's LADEE mission. The spacecraft orbits 250 km (155 miles) above the surface of the Moon, searching for dust in the lunar atmosphere.

Solar Storms Sandblast the Moon's Surface

Solar storms and associated coronal mass ejections (CMEs) can significantly erode the lunar surface, according to a new set of computer simulations by NASA scientists. In addition to removing a

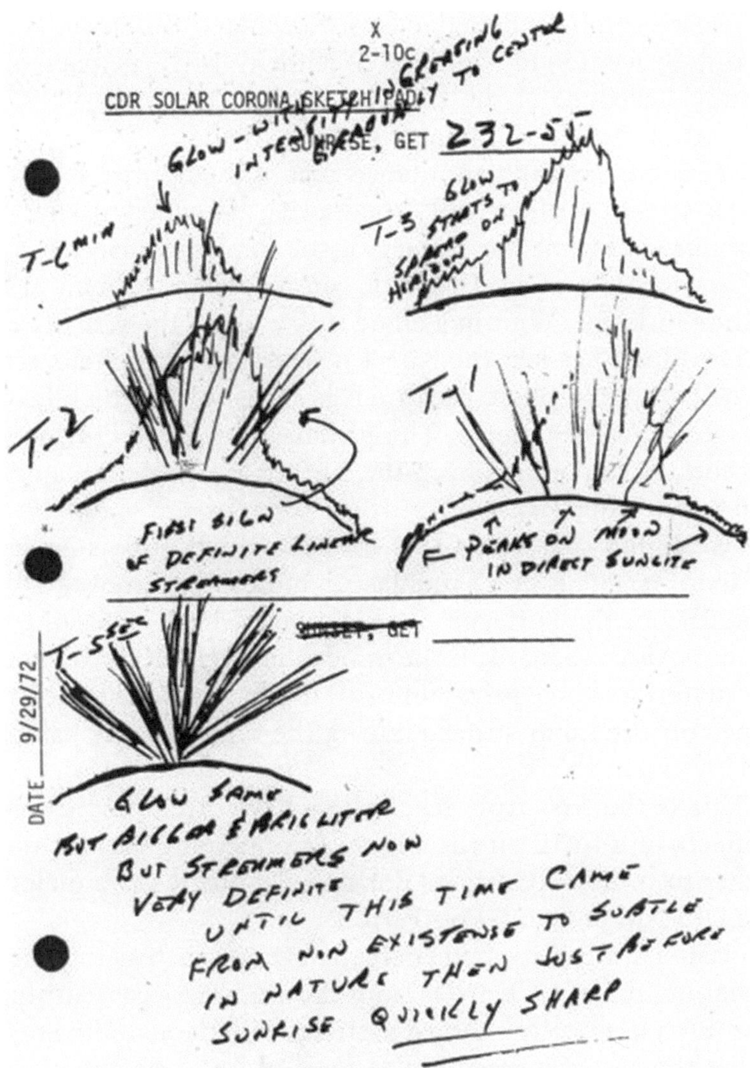

FIG. 9.5 *Apollo 17* astronauts orbiting the Moon in 1972 repeatedly saw and sketched what they variously called "bands," "streamers," or "twilight rays" for about 10 s before lunar sunrise or lunar sunset. Such rays were also reported by astronauts aboard *Apollo 8, 10,* and *15*. NASA's LADEE spacecraft also found evidence of this levitating dust in the lunar atmosphere. This is similar to crepuscular rays on Earth (Credit: NASA)

surprisingly large amount of material from the lunar surface, this could be a major method of atmospheric loss for planets like Mars that are unprotected by a global magnetic field.

The research is being led by Rosemary Killen at NASA's Goddard Space Flight Center, Greenbelt, MD, as part of the Dynamic Response of the Environment At the Moon (DREAM) team within the NASA Lunar Science Institute.

CMEs are basically an intense gust of the normal solar wind, a diffuse stream of electrically conductive gas called plasma that's blown outward from the surface of the Sun into space. A strong CME may contain around a billion tons of plasma moving at up to a million miles per hour in a cloud many times the size of Earth.

The Moon has just the barest wisp of an atmosphere, technically called an exosphere because it is so tenuous, which leaves it vulnerable to CME effects. The plasma from CMEs impacts the lunar surface, and atoms from the surface are ejected in a process called "sputtering."

"We found that when this massive cloud of plasma strikes the Moon, it acts like a sandblaster and easily removes volatile material from the surface," said William Farrell, DREAM team lead at NASA Goddard. "The model predicts 100–200 tons of lunar material—the equivalent of 10 dump truck loads—could be stripped off the lunar surface during the typical 2-day passage of a CME."

This is the first time researchers have attempted to predict the effects of a CME on the Moon. "Connecting various models together to mimic conditions during solar storms is a major goal of the DREAM project," says Farrell.

Plasma is created when energetic events, such as intense heat or radiation, remove electrons from the atoms in a gas, turning the atoms into electrically charged particles called ions. The Sun is so hot that the gas is emitted in the form of free ions and electrons called the solar wind plasma. Ejection of atoms from a surface or an atmosphere by plasma ions is called sputtering.

"Sputtering is among the top five processes that create the Moon's exosphere under normal solar conditions, but our model predicts that during a CME, it becomes the dominant method by far, with up to 50 times the yield of the other methods," says Killen, lead author of a paper on this research appearing in a special issue of the *Journal of Geophysical Research Planets*.

CMEs are effective at removing lunar material not only because they are denser and faster than the normal solar wind but also because they are enriched in highly charged, heavy ions, according to the team. The typical solar wind is dominated by lightweight hydrogen ions (protons). However, a heavier helium ion with more electrons removed, and hence a greater electrical charge, can sputter tens of times more atoms from the lunar surface than a hydrogen ion.

The team used data from satellite observations that revealed this enrichment as input to their model. For example, helium ions comprise about 4 % of the normal solar wind, but observations reveal that during a CME, they can increase to over 20 %. When this enrichment is combined with the increased density and velocity of a CME, the highly charged, heavy ions in CMEs can sputter 50 times more material than protons in the normal solar wind.

"The computer models isolate the contributions from sputtering and other processes," says Dana Hurley, a co-author on the paper at the Johns Hopkins University Applied Physics Laboratory in Laurel, MD. "Comparing model predictions through a range of solar wind conditions allows us to predict the conditions when sputtering should dominate over the other processes. Those predictions can later be compared to data during a solar storm."

NASA's LADEE, a lunar orbiter mission launched in 2013—was able to test their predictions. The strong sputtering effect kicked lunar surface atoms to LADEE's orbital altitude, around 20–50 km (12.4–31 miles), so the spacecraft saw them increase in abundance.

"This huge CME sputtering effect made LADEE almost like a surface mineralogy explorer, not because LADEE is on the surface but because during solar storms surface atoms are blasted up to LADEE," said Farrell.

The Moon is not the only heavenly body affected by the dense CME driver gas. Space scientists have long been aware that these solar storms dramatically affect Earth's magnetic field and are responsible for intense aurorae (the Northern and Southern Lights).

On exposed small bodies such as asteroids, the dense, fast-streaming CME gas should create a sputtered-enhanced exosphere about the object, similar to that expected at the Moon.

Crater Rays Point to New Lunar Time Scale

The lunar time scale should be reevaluated, suggest remote sensing studies of lunar crater rays by B. Ray Hawke (University of Hawai'i) and colleagues at the University of Hawai'i, NovaSol, Cornell University, National Air and Space Museum, and Northwestern University. These scientists have found that the mere presence of crater rays is not a reliable indicator that the crater is young, as was once thought, and that the working definition of the Copernican/Eratosthenian (C/E) boundary should be reconsidered.

The team used Earth-based spectral and radar data with FeO, TiO_2, and optical maturity maps derived from Clementine UVVIS images to determine the origin and composition of selected lunar ray segments. They conclude that the optical maturity parameter, which uses chemical analyses of lunar samples as its foundation, should be used to redefine the C/E boundary (Fig. 9.6).

Under this classification, the Copernican system would be defined as the time required for an immature surface to reach full optical maturity. Lunar crater rays are those obvious bright streaks of material that we can see extending radially away from many impact craters. Historically, they were once regarded as salt deposits from evaporated water (early 1900s) and volcanic ash or dust streaks (late 1940s).

Beginning in the 1960s, with the pioneering work of Eugene Shoemaker, rays were recognized as fragmental material ejected from primary and secondary craters during impact events. Their formation was an important mechanism for moving rocks around the lunar surface, and rays were considered when planning the Apollo landing sites. A ray from Copernicus Crater crosses the *Apollo 12* site in Oceanus Procellarum. North Ray and South Ray craters have rays that cross near the *Apollo 16* site in the Descartes Highlands, and a ray from Tycho Crater can be traced across the *Apollo 17* site in the Taurus-Littrow Valley on the eastern edge of Mare Serenitatis. There is still much debate over how much ejecta comes from the primary impact site or by secondary craters that mix local bedrock into ray material.

To sort out the nature of the rays, Hawke and his colleagues focused their study on rays associated with four craters: Tycho,

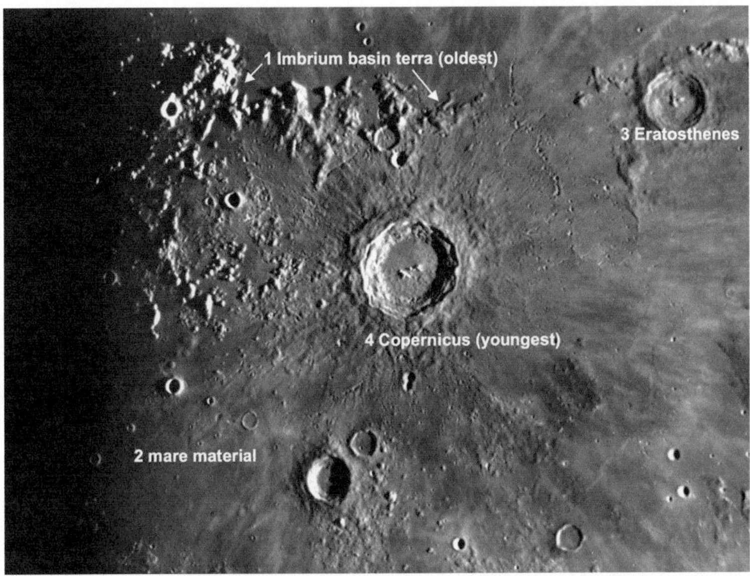

FIG. 9.6 This is a telescopic image of Copernicus and vicinity, showing how the relative ages of geological features are determined using the principle of superposition. This states that the oldest rocks are those that form the highland units of the large, circular Imbrium impact basin. These units are the mountains that make up the rim of the basin as well as the regional highlands around Copernicus, which are ejecta from the basin forming event. Partial flooding by the dark, smooth maria followed, including both dark ash-like materials and smooth flood-like plains. These eruptions were followed by the formation of impact craters, of which two kinds could be recognized: an older group that had slightly eroded and lost its bright rays (such as Eratosthenes) and a younger group that preserved the bright rays and showed a fresh, unmodified form (such as Copernicus) (Credit: NASA)

the Messier Crater complex, Lichtenberg, and Olbers A. They combined Earth-based observations with Clementine-derived maps to investigate ray compositions, maturities, and modes of origin, and to assess the consequence of their findings on the lunar time scale. Chemical analyses of lunar samples provide "ground truth" to understand the geologic processes on the Moon. Specifically, the discoveries by cosmochemists of nanophase-iron grain coatings, Fe-Si phases, and other space weathering products in lunar rocks and meteorites have enabled us to better understand the physical, chemical, and optical changes that occur over time as the lunar surface is exposed to the space environment and matures. Older surfaces in which these changes have reached a steady state are said to be fully mature. Younger surfaces are called immature.

Space weathering products in the lunar material, which can only be discovered by sample analysis, affect the spectral signatures of the Moon's surface. Detailed studies of the variation of spectra with extent of space weathering have been made by Larry Taylor (University of Tennessee, Knoxville), Carlé Pieters (Brown University) and their colleagues. Using scanning electron microscopy, they've measured, for example, the abundance of agglutinate glass and major minerals in grain-size separates of lunar regolith. Three types of data were compiled and analyzed by the research team to study the lunar rays.

Maps of FeO, TiO_2, and optical maturity were derived from Clementine UVVIS images (415–1000 nm) and sample analyses were done, using the methods developed by Paul Lucey (University of Hawai'i) and co-workers. Spatial resolution was ~100 m/pixel.

Near-IR reflectance spectra between 0.6 and 2.5 μm (600–2500 nm) were collected at the University of Hawai'i 2.24-m telescope at Mauna Kea Observatory. Spatial resolutions (called spot sizes) were ~1.5 and 4.5 km in diameter. Spectral curves showed absorption bands near 1 μm characteristic of iron-bearing silicate minerals. They used the shape and position of this band to determine the composition and relative abundance of pyroxene and olivine minerals, which helped them distinguish rock types.

Radar data at three wavelengths—3.0, 3.8, and 70 cm—were transmitted and received by radar antennae at Haystack Observatory (Massachusetts) and Arecibo Observatory (Puerto Rico).

Surface and subsurface scattering properties of the Moon were analyzed using these radar backscatter images. The 3.0 and 3.8 cm radar data are sensitive to roughness on the scales of 1–50 cm within the upper meter of the regolith. The 70 cm data show roughness on scales from 50 cm to 10 m within 5–10-m depths. Rough (blocky) ray deposits are bright in radar images. As these young rays are exposed to space weathering, they mature and become smoother so they appear darker in the radar images.

General point to consider: Bright areas in the albedo images are either due to the presence of low FeO material or an immature surface. The dark lunar maria are lava plains composed of a Fe-rich rock called basalt. The brighter highlands are made of rocks much lower in FeO. Near-IR spectra as well as the FeO, TiO_2, and maturity maps indicate that the southern and western rays of the Messier complex are composed of debris from immature mare basalts. They appear bright in radar images. Highlands material is not present in these rays; hence they are bright because they are immature.

A portion of the Tycho ray in Mare Nectaris is composed largely of immature mare basalt with little to no detectable Tycho ejecta material. The ray is dominated by fresh local material excavated and emplaced by the secondary craters as well as fresh material that is constantly exposed on the crater walls due to landslides. This ray segment is bright because it is immature.

Tycho Ray in Southern Highlands is composed of relatively immature highland debris. But it is not possible to determine how much of the material is local and how much is projectile material blasted in from Tycho Crater. This ray segment is bright because it is immature.

FeO and TiO_2 abundances for Lichtenberg Crater rays are consistent with highland rocks. The optical maturity map demonstrates that these highlands-rich ejecta deposits and rays are fully mature. Hence, these rays are bright because of their composition.

This high-albedo ray for Olbers A Ray, which was deposited on a mare surface, has reduced FeO and TiO_2 abundances, consistent with the presence of a large non-mare component. Much of the ray is not distinct in the optical maturity map, but some areas are bright in the OMAT (see arrows A, B, and C) suggesting that these areas are not as mature as the adjacent terrain. This ray has a significant amount of highlands ejecta debris and is bright because of composition and immaturity, and is a good example of a "combination" ray.

The work by Hawke and others shows that the brightness of rays is due to immaturity and/or compositional differences. There are craters with compositional rays whose ages are significantly greater than 1.1 billion years. Lichtenberg is a prime example. This 20-km-diameter crater is embayed on the southeast by mare basalt flows. Harry Hiesinger (Brown University) and his coworkers have estimated the age of these flows on the basis of the number of craters formed on them. They report an age of 1.6–2 billion years, distinctly older than the nominal C/E boundary age of 1.1 billion years. It follows that the mere presence of rays is not a reliable indicator of crater age. And it is no longer valid to assign a Copernican age to craters based only on the presence of rays.

Hawke and others conclude that a new method using the optical maturity parameter is required to distinguish Copernican from Eratosthenian craters. They acknowledge a problem of not knowing the time required for a surface to reach full optical maturity; no such age has been firmly established.

A possible solution was proposed by Jennifer Grier (formerly at the University of Arizona and now at the Harvard-Smithsonian Center for Astrophysics) and Al McEwen (University of Arizona) and colleagues. Their work showed that if the ejecta of the Copernicus crater were slightly more mature, it would be impossible to tell apart from the optically mature bedrock. Since the commonly accepted age of Copernicus is about 0.8 billion years, then perhaps full optical maturity occurs at about 0.8 billion years.

More work is necessary and future studies will look more closely at optical maturity maps of the Copernicus crater region to better define the C/E boundary in the lunar time scale. One particularly useful way to pin down the ages of specific craters is to determine their absolute ages by isotopically dating samples returned from them. Cosmochemists could date either samples of impact melt from the floors of the craters or samples of mare basalts that embay ejecta from the craters, as at Lichtenberg.

10. More Moon Mysteries

Is the Moon a Planet?

The controversy over what makes a celestial body a planet has been going on for decades. In fact the planet Pluto became a casualty of that debate when it was demoted to a planetoid. So if Pluto can get the axe as a planet, then surely the Moon can be eventually promoted to being a planet. However we have to go by the currently accepted understanding of what a planet is.

According to the International Astronomical Union (IAU) a planet is an object that orbits the Sun and is large enough to have become round due to the force of its own gravity. In addition, a planet has to dominate the neighborhood around its orbit. Pluto has been demoted because it does not dominate its neighborhood. Charon, its large "moon," is about half the size of Pluto, while all the true planets are far larger than their Moons. In addition, planets must dominate their neighborhoods—"sweep up" asteroids, comets, and other debris, clearing a path along their orbits. By contrast, Pluto's orbit is somewhat untidy.

A new wrinkle in the definition of a planet could be just the thing to put the Moon over the top in the future. According to the current definition a moon can't become a planet if it orbits another planet; however it can be considered one if the center of gravity, known as the barycenter, lies outside of the larger planet. Right now the Moon is still solidly in the natural satellite category, but each year it is drifting further and further away in its orbit from Earth. It's currently drifting away about 1.5 in. (3.74 cm) every year. For now, the system's barycenter is inside Earth. But that will change. This means that someday in the distant future scientists may consider reclassifying it as a planet. That is, however, if the definition for what is a planet does not change in that time (Fig. 10.1).

None of this would occur for a few billion years. And Earth and the Moon would have to survive a host of remote catastrophe

© Springer International Publishing Switzerland 2016

V.S. Foster, *Modern Mysteries of the Moon*, Astronomers' Universe,

DOI 10.1007/978-3-319-22120-5_10

FIG. 10.1 When the barycenter moves outside the Earth, the Moon may become a planet (Credit: NASA)

scenarios along with the predicted swelling of the Sun into a red giant, which might engulf and vaporize our planet (unless we can figure out a way to move it).

Astronomers have noted that it is possible there are three-object systems yet to be found in the outer Solar System. If they are all round and have that certain barycenter thing happening, then they'd be called triple planets under the new definition.

Astronomers expect to find hundreds of Pluto-sized objects in the outer Solar System. If one has a satellite that is round, and which has a certain eccentric orbit—meaning the two objects come very close together at one point and then diverge greatly—then the barycenter could dip inside the larger object during part of the orbit. In such a case, the smaller object would be defined as a moon part of the time and a planet the rest.

The Earth-Moon system may be considered a double, or binary, planet, informal terms used to describe a binary system where both objects are of planetary mass. Though not an official classification, the European Space Agency has referred to the Earth-Moon system as a double planet. The IAU General Assembly in August 2006 considered a proposal that Pluto and Charon be reclassified as a double planet, but the proposal was abandoned.

There has been debate on where to draw the line between a double planet and a planet-moon system. In most cases, this is not an issue because most satellites have small masses relative to their planets. In particular, with the exception of the Earth-Moon system, all satellites of planets in the Solar System have masses less than 0.00025 (1/4000) the mass of the host planet. The Moon-to-Earth mass ratio is 0.01230 ($\approx$1/81). In comparison, the Charon-to-Pluto mass ratio is 0.117 ($\approx$1/9).

The now-abandoned co-accretion hypothesis of the origin of the Moon is also called the double-planet hypothesis. The idea is that two bodies should be considered a double planet if they accreted together directly from the protoplanetary disk, much as a double star typically forms together.

Once it was realized that both Earth's Moon and Pluto's Charon likely formed from giant impacts, this parallel was noted when calling them double planets. However, an impact may also produce tiny satellites, such as the small outer satellites of Pluto, so this does not determine where the line should be drawn.

Isaac Asimov suggested a distinction between planet-moon and double-planet structures based in part on what he called a "tug-of-war" value, which does not consider their relative sizes. This quantity is simply the relationships between the masses of the primary planet and the Sun combined with the squared distances between the smaller object and its planet and the Sun:

$$\text{tug-of-war value} = \frac{m_1}{m_2} \times \left(\frac{d_1}{d_2}\right)^2$$

Where m_1 is the mass of the larger body, m_2 is the mass of the Sun, d_1 is the distance between the smaller body and the Sun, and

d_2 is the distance between the smaller body and the larger body. Note that the tug-of-war value does not rely on the mass of the satellite or smaller body.

This formula actually reflects the relation of the gravitational effects on the smaller body from the larger body and from the Sun. The tug-of-war figure for Saturn's moon Titan is 380, which means that Saturn's hold on Titan is 380 times as strong as the Sun's hold on Titan. Titan's tug-of-war value may be compared with that of Saturn's moon Phoebe, which has a tug-of-war value of just 3.5. So Saturn's hold on Phoebe is only 3.5 times as strong as the Sun's hold on Phoebe.

Asimov calculated tug-of-war values for several satellites of the planets. He showed that even the largest gas giant, Jupiter, had only a slightly better hold than the Sun on its outer captured satellites, some with tug-of-war values not much higher than one. Yet in nearly every case the tug-of-war value was found to be greater than one, so in every case the Sun loses the tug-of-war with the planets. The one exception was Earth's Moon, where the Sun wins the tug-of-war with a value of 0.46, which means that Earth's hold on the Moon is less than half the Sun's hold. Since the Sun's gravitational effect on the Moon is more than twice that of Earth's, Asimov reasoned that Earth and the Moon form a binary planet structure. This was one of several arguments in Asimov's writings for considering the Moon a planet rather than a satellite. Asimov said:

We might look upon the Moon, then, as neither a true satellite of the Earth nor a captured one, but as a planet in its own right, moving about the Sun in careful step with the Earth. From within the Earth-Moon system, the simplest way of picturing the situation is to have the Moon revolve about the Earth; but if you were to draw a picture of the orbits of the Earth and Moon about the Sun exactly to scale, you would see that the Moon's orbit is everywhere concave toward the Sun. It is always "falling toward" the Sun. All the other satellites, without exception, "fall away" from the Sun through part of their orbits, caught as they are by the superior pull of their primary planets—but not the Moon.

Note that this definition of "double planet" depends primarily on the two-body structure's distance from the Sun. If the Earth-Moon system happened to orbit farther away from the Sun than it does now, then Earth would win the tug of war. From the orbit of Mars, the Moon's tug-of-war value would be 1.05, so the Sun would no longer win the tug of war with Earth. Also, several tiny moons discovered since Asimov's argument would qualify as double planets.

What If Earth Had No Moon?

Can you imagine Earth without a Moon? No beautiful, bright object traversing the night sky, hovering on the horizon, peeking through the trees on a cold winter's night? No romantic moonlight, no song *Blue Moon*, and no lunar landings.

Not only would we miss the Moon. Without the Moon, we might not even exist.

What would have happened on Earth if Theia had passed peacefully on its way without striking Earth and forming the Moon? Well, life of some sort would probably exist on Earth, but humans almost certainly wouldn't. Think of the very long course of evolution, the small changes, the minute adaptations that organisms make to their environment. It would only have taken small changes to Earth's environment to have dramatically altered the course of evolution. I probably wouldn't be writing this book—and you probably wouldn't be reading it.

The Moon's largest effect on Earth is the tide. According to the law of universal gravitation, any two objects in the universe pull each other, and the force of this pull is in direct proportion to the objects' masses and in inverse proportion to the square of the distance between the objects. The gravitational pull between Earth and the Moon causes the seas and the oceans on Earth to either rise or subside. The effect is called the tide, which changes between high tide and low tide, according to the Moon's position. One third of all tide effects on Earth are caused by the Sun's pull and the rest is caused by the Moon's (Fig. 10.2).

When life first formed on Earth in tidal pools, it was thanks to the gravitational pull of the Moon that primordial life was able to traverse between different pools and generally spread across the planet. Without the Moon, the churning of the oceans, and thus the circulation of nutrients, ceases. Water-based life struggles to survive and, eventually, thousands (and probably millions) of species go extinct.

Fig. 10.2 Residents around Canada's Bay of Fundy are among the Earthlings who are most affected by the Moon's influence. They cope with a difference of up to 15 m between high and low tides (Credit: NASA)

Earth did not always have a Moon, so where did it come from? As discussed earlier, the leading scientific theory is that an object about the size of Mars, called Theia, collided with Earth about 4.5 billion years ago. Striking at an oblique angle, it raised a cloud of debris that then coalesced to form the Moon. This had profound effects on Earth.

And if the Moon had never formed, Earth would be a very, very different place. Imagine a Moon-less weather report—blizzards over the Sahara, floodwaters swallowing the Pyramids, 90° temperatures in Antarctica. As Earth wobbles on its axis—unsecured by the Moon's gravitational pull—the polar caps would grow and recede at frightening rates. And without the Moon, our planet would spin much faster—meaning shorter days and searing temperatures.

An Earth day would be only 8–10 h long, with no Moon to slow it down. The faster rotation would cause winds of 160–200 km to sweep Earth's surface. The tilt axis of Earth would wobble, resulting in dramatic changes in temperature over thousands to millions of years. And although our seas would still be tidal, the tides would be much smaller—caused only by the Sun.

Earth and its newly formed Moon exerted a gravitational force on each other, slowing the rotation of Earth and lengthening the Earth day from 5 to 24 h. In fact, to this day, the Moon continues to slow down the rotation of Earth, although only by 0.002 s/century.

The gravitational attraction between Earth and the Moon also stabilized the tilt of Earth's axis, and it is today's constant tilt of 23.5° that gives Earth its predictable, fairly constant climate and its seasons. Without the Moon, however, the axis would have continued to wobble, as does that of Planet Mars (Fig. 10.3).

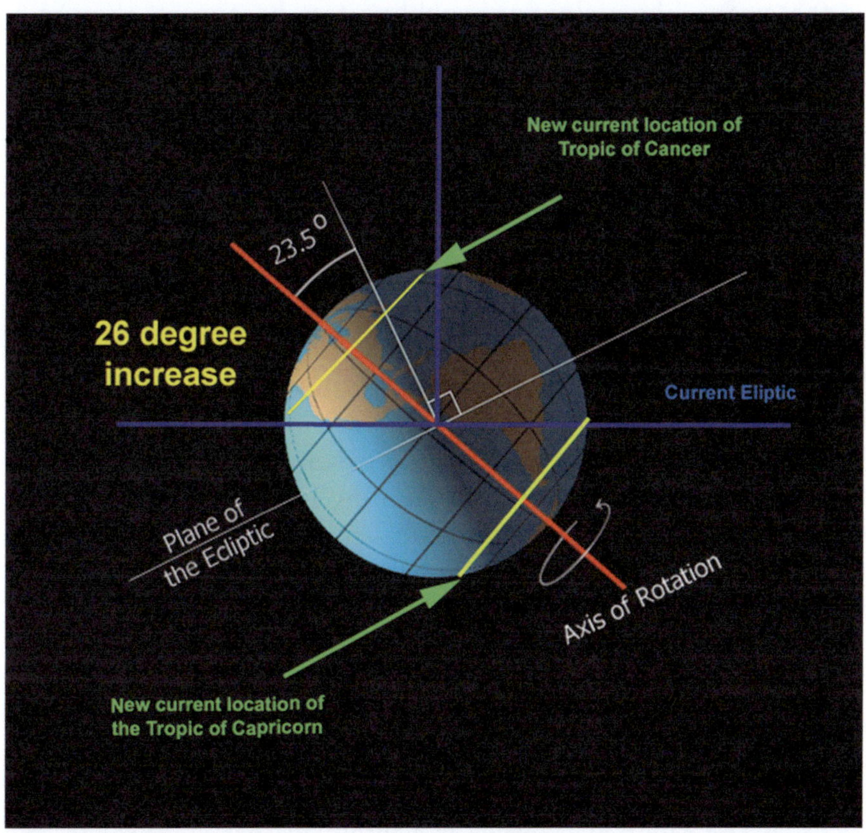

FIG. 10.3 Currently Earth's axial tilt is 23.4°, but it fluctuates between the two angles in the illustration in a 41,000-year period. Without the Moon it would tilt much more (Credit: NASA)

With no moon as a stabilizer, the Earth would sometimes tilt all the way over and lie on its side in relation to its orbit around the Sun. This would make for extreme differences between temperatures and daylight throughout the year.

At other times Earth's axis would be straight up and down, making night and day equally long, year round, and there would be no seasons.

There would be periods with more extreme weather, and bigger differences between winter and summer.

"A good example of this is Mars, which has no large moon to stabilize it, so it tilts more. The Martian climate and atmosphere has undergone enormous changes in the past millions of years. We probably would see something similar to that here," says Aksnes.

Right now the spin axis of Earth very slowly wobbles over 26,000 years, like a slowing wobbling top, because of the tug of the Sun. The wobble causes true north to not always point at Polaris, a.k.a., the North Star. Experts agree that the Moon acts sort of like a shock absorber to this wobble—keeping it from getting out of hand.

Without the Moon, Earth would have spun faster, causing a faster heat exchange among the air, seas, and land. This, in turn, would have caused hurricanes in the east–west direction on Earth with speeds up to 100 miles/h. Such conditions would have been very problematic to all complex life forms, including human beings. Even simple tasks such as speaking or listening could have been very difficult or even impossible. Since a day would have been 8 h, this would have caused a mismatch between the biological clock of living things—including human beings—and the flow of the day, resulting in biological complications. Without the Moon, the high tide would have been very weak, again causing a very inconvenient environment for sea creatures.

The Moon also has a role in keeping Earth's axis at a 23.5° angle. The seasons result from this, with just the right amount of sunlight going to the equator and the poles to create a climate appropriate for life. Another known effect of the Moon on Earth is that the Moon reflects the Sun's light and heats up Earth by 32.3 °F (0.2 °C). The Moon also serves as a shield against space rocks, and if it had not been for the Moon many more rocks and meteors would have hit Earth.

It's possible that Earth without a Moon would wobble wildly, sort of like Mars does. The Red Planet's wobble is so extreme that it may be the cause of some cycles of climate change there. If the same thing happened here, Earth might wobble so much that seasons would become inhospitably extreme, and Earth would be a much less stable and habitable planet.

Without the Moon the tilt of Earth's axis could go from its current wobble of 22°–25° to a wide ride of 0°–85°. Zero would eliminate seasons, and 85 is basically Earth leaning over on its side. If this happened, the current crisis we call global warming would be a very pleasant tea party by comparison.

Luckily, the wobbling would not affect things right away but over many millions of years.

Without the Moon, the stability of Earth's axis would be lost, and with it, our predictable temperatures. Let us consider two cities: Rome, Italy, and Stockholm, Sweden. In summer the average high temperature in Rome is 29 °C, and in the winter the average high temperature is 13 °C. In Stockholm, the high in summer is 20 °C and in winter it is 0 °C. If Earth's tilt axis changed, the temperatures in these two cities would change dramatically. Imagine if the temperatures were swapped: the infrastructure (e.g., air conditioning or snow ploughs) would simply not be in place in those cities for humans to live, work, and eat comfortably. The Italians, Swedes, and all other life on Earth would need to adapt or face extinction.

Moving might be one option, but not for all organisms. Coral reefs, for example, are sensitive, and complex ecosystems that might not be able to adapt fast enough to the changing water temperature and would probably die.

Another feature of our planet is its oceans. More than 70 % of Earth's surface is covered by salt water, rising and falling on a 12.5-h tidal cycle. The forces that create tides are complex, involving not only the centrifugal forces of Earth's rotation but also the gravitational pull of both the Moon and the Sun. The effect of the Moon, however, is twice that of the Sun; this is because the gravitational force that one object exerts on another depends on both its mass and its distance.

With the lack of stability in Earth's tilt, we would also lose our regular seasons with far-reaching consequences. Think how many organisms grow, mate, migrate, or hibernate at particular times of the year. And drastic changes in temperature would affect the growing season and climate for plants, making food production for the billions of people on Earth more complex.

What kinds of life might have evolved on a moonless Earth, capable of withstanding extreme temperatures, high winds, small tides, and short days? Well, life of some sort would probably exist on Earth, but humans as we know them almost certainly wouldn't. Think of the very long course of evolution, the small changes, the minute adaptations that organisms made to their environment. It only would have taken a small change to Earth's environment to have dramatically altered the course of evolution.

We do not know how close the Moon was to Earth when it first formed, but we do know that it was farther than 12,000 km and closer than it is today (about 384,400 km). This means that it initially caused much larger tides than we experience today—tides that are thought to have been important in mixing the oceans and in the early evolution of life, some 3.8 billion years ago.

Most of the cosmic rays that come from space are neutralized by Earth's magnetic field. Few of these rays reach Earth and cause chemical reactions. Without the Moon, Earth' core would have spun faster along with Earth itself. With the core of Earth spinning faster, the magnetic field would have been much stronger. This would have caused huge changes in the atmosphere. Besides, some bacteria and animals that use the magnetic field to find their way (such as sea turtles, salmon, eels, pigeons, and migratory birds) would have been negatively affected. Consequently many ecosystems, as we know them today, would have been different.

Interestingly, the tides and the rotation of Earth have an effect on the Moon. Together, they pull on the Moon, making it spin just a little faster; and as it spins faster and faster, it moves further away from Earth—albeit at a rate of only 3.82 cm/year.

Imagine that there were no solar or lunar eclipse. This would have changed our early views of the shape of Earth. During a lunar eclipse the curved shadow of our planet can be seen moving across the face of the Moon, giving away the round shape of Earth.

Many of our vocabulary words are "Moon oriented." The words "honeymoon," "lunatic," "moonshine," and "moonbeam" would not exist. And try to imagine us without that showy social past-time of "mooning!"

Earth certainly would have been a different place.

A Hollow Moon?

Why does the Moon ring like a hollow sphere when a large object hits it? During the Apollo missions, ascent stages of lunar modules as well as the spent third stages of rockets crashed on the surface of the Moon. Each time, these caused the Moon to "ring like a gong or a bell," according to NASA. The reverberations lasted for over an hour.

To uncover the answer to this puzzle, it took planetary geologists over four decades to collect and study moonquake data and develop a clear understanding of the internal structure of the Moon.

What caused these reverberations was an "echo chamber" in the interior of the Moon. Scientists theorize that the echo chamber is a jumbled layer of rocky debris sandwiched between the hard lunar surface and underlying hard material. The debris may have been formed eons ago during some cataclysmic lunar event. For some unknown reason, the theorists say, energy from a crash bounces back and forth through the rubble like sound in an echo chamber. The Moon's density is 3.34 g/cm^3 (3.34 times an equal volume of water) whereas Earth's is 5.5. Proponents argue that this indicates the Moon must have a large cavity inside it.

Seismic devices left by the astronauts detected the reverberations, which were then radioed to Earth. The Moon is relatively dry and has no fluids trapped in it, as Earth does, to dampen energy reverberations from crashes on the surface (Fig. 10.4).

Moonquake data collected since the Apollo era helped scientists understand what lay beneath the Moon's crust. The research disclosed that the Moon possesses a solid, iron-rich inner core

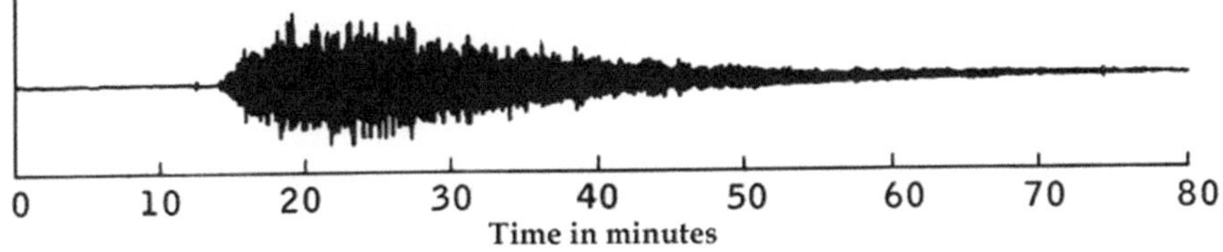

LM impact measured at the Apollo 12 Seismometer.

FIG. 10.4 Seismogram recording the impact of *Apollo 12* ascent stage. NASA

with a radius of nearly 150 miles and a fluid, primarily liquid-iron outer core with a radius of roughly 205 miles.

Where it differs from Earth is a partially molten boundary layer around the core, estimated to have a radius of nearly 300 miles. The data sheds light on the evolution of a lunar dynamo—a natural process by which our Moon may have generated and maintained its own strong magnetic field.

Uncovering details about the lunar core is critical for developing accurate models of the Moon's formation. The core contains a small percentage of light elements such as sulfur, echoing new seismology research on Earth that suggests the presence of light elements—such as sulfur and oxygen—in a layer around our own core. The research used extensive data gathered from the Apollo Passive Seismic Experiment (APSE), consisting of four seismometers deployed between 1969 and 1972, which recorded continuous lunar seismic activity until late 1977, when they were turned off in a cost-cutting move.

The Moon is a differentiated body, being composed of a geochemically distinct crust, mantle, and core. This structure is believed to have resulted from the fractional crystallization of a magma ocean shortly after its formation about 4.5 billion years ago. The energy required to melt the outer portion of the Moon is commonly attributed to a giant impact event that is postulated to have formed the Earth-Moon system, and the subsequent reaccretion of material in Earth orbit.

Geochemical mapping from orbit shows that the crust of the Moon is largely anorthositic in composition, consistent with the magma ocean hypothesis. In terms of elements, the lunar crust is composed primarily of oxygen, silicon, magnesium, iron, calcium, and aluminium, along with important minor and trace elements of titanium, uranium, thorium, potassium, and hydrogen. The crust is estimated to be on average about 50 km thick.

Partial melting within the mantle of the Moon gave rise to the eruption of mare basalts on the lunar surface. Analyses of these basalts indicate that the mantle is composed primarily of the minerals olivine, orthopyroxene, and clinopyroxene, and that the lunar mantle is richer in iron than Earth is. Some lunar basalts contain high abundances of titanium (present in the mineral ilmenite), suggesting that the mantle is highly heterogeneous in composition.

Moonquakes have been found to occur deep within the mantle of the Moon about 1000 km below the surface. These occur with monthly periodicities due to tidal stresses caused by the eccentric orbit of the Moon around Earth. A few shallow moonquakes with hypocenters located about 100 km below the surface have also been detected, but these occur more infrequently and appear to be unrelated to the lunar tides.

Several lines of evidence indicate that the lunar core is small, with a radius of about 350 km or less. The size of the lunar core is only about 20 % the size of Earth, in contrast to about 50 %, as is the case for most other terrestrial bodies. The composition of the lunar core is not well known, but most believe that it is composed of metallic iron alloyed with a small amount of sulfur and nickel. Analyses of the Moon's time-variable rotation indicate that the core is at least partly molten.

Reanalysis of the old Apollo seismic data on the deep moonquakes using modern processing methods in 2010 confirmed that the Moon has an iron-rich core with the radius of 330 ± 20 km. The same reanalysis established that the solid inner core made of pure iron has the radius of 240 ± 10 km. The core is surrounded by the partially (10–30 %) melted layer of the lower mantle with the radius of 480 ± 20 km (thickness ~150 km). These results imply that 40 % of the core by volume has solidified. The density of the liquid outer core is about 5 g/cm^3 and it can contain as much 6 % sulfur by weight. The temperature of the core is probably about 1600–1700 K.

The evidence indicates that the Moon was active geologically in its early history, but it has been essentially dead geologically for more than 3 billion years. Based on that evidence, we believe the chronology of lunar geology is as follows:

1. The Moon was formed about 4.6 billion years ago. The surface was subjected continuously to an intense meteor bombardment from debris left over from the formation of the Solar System.
2. By about 4.4 billion years ago the top 100 km was molten, from original heat of formation and from heat generated by the meteor bombardment.
3. By 4.2 billion years ago the surface was solid again.
4. As the intense meteor bombardment associated with debris left over from the formation of the Solar System continued, most of

the craters that we now see on the surface of the Moon were formed by meteor impact.

5. Fracturing and heating of the surface and subsurface by the meteor bombardment led to a period of intense volcanic activity 3.8–3.1 billion years ago. Meanwhile, the meteor bombardment had tapered off because by this time much of the debris of the early Solar System had already been captured by the planets.

6. Lava flows filled the low areas and many craters. These flows solidified to become the flat and dark maria, which have little cratering because most of the original craters were covered by lava flows and only a few meteors of significant size have struck the surface since the period of volcanic activity. The regions that were not covered by the lava flows are the present highlands; thus, they are heavily cratered, and formed from different rocks than the seas.

7. Volcanism stopped about 3.1 billion years ago. The Moon has been largely dead geologically since then except for the occasional meteor impact or small moonquake, and micrometeorite erosion of the surface.

Thus, lunar surface features, particularly in the highlands, tend to be older than those of Earth, which remains to this day a geologically active body.

Earth's Second Moon

A new theory suggests that Earth once had a small second moon that perished in a slow motion collision with its "big brother." Researchers suggest the collision may explain the mysterious mountains on the far side of our Moon. The smaller 'twin' Moon is believed to have only survived a few million years before it collided with the one we see today, leaving just one. The Moon's smaller twin is believed to have been about one-thirtieth of the size. It would have orbited Earth at the same speed and distance and just got slowly sucked in until they hit and then coalesced.

Earth and its present Moon are thought to have been formed between 30 million and 130 million years after the birth of the Solar System, about 4.6 billion years ago. The second moon would have lasted for only a few million years; then it would have collided with the Moon to leave the one large body we see today.

Scientists note that the relatively slow speed of the crash was crucial in adding material to the rarely seen lunar hemisphere. Details have been published in the journal *Nature*.

For decades, scientists have been trying to understand why the near side of the Moon—the one visible from Earth—is flat and cratered while the rarely seen far side is heavily cratered and has mountain ranges higher than 3000 m.

Various theories have been proposed to explain what's termed the lunar dichotomy. One suggests that tidal heating, caused by the pull of Earth on the ocean of liquid rock that once flowed under the lunar crust, may have been the cause.

But a recent study proposes a different solution: a long-term series of cosmic collisions. The researchers argue that Earth was struck about 4 billion years ago by another planet about the size of Mars. This is known as the global-impact hypothesis. The resulting debris eventually coalesced to form our Moon. Scientists say that another, smaller lunar body may have formed from the same material and become stuck in a gravitational tug of war between Earth and the Moon.

Dr. Martin Jutzi from the University of Bern, Switzerland, is one of the authors of the paper. He explained: "When we look at the current theory there is no real reason why there was only one Moon. And one outcome of our research is that the new theory goes very well with the global impact idea."

After spending millions of years "stuck," the smaller moon embarked on a collision course with its big brother, slowly crashing into it at a velocity of less than 3 km/s—slower than the speed of sound in rocks.

Jutzi says it was a low velocity crash: "It was a rather gentle collision, at around 2.4 km/s, lower than the speed of sound—that's important because it means no huge shocks or melting was produced. At the time of the smash, the bigger Moon would have had a "magma ocean" with a thin crust on top."

The scientists argue that the impact would have led to the build-up of material on the lunar crust and would also have redistributed the underlying magma to the near side of the Moon, an idea backed up by observations from NASA's Lunar Prospector spacecraft.

In a commentary Dr. Maria Zuber, from the Massachusetts Institute of Technology (MIT) in Cambridge, suggests that while the new study "demonstrates plausibility rather than proof," the authors "raise the legitimate possibility that after the giant impact our Earth perhaps fleetingly possessed more than one Moon."

The researchers believe one way of proving their theory is to get their hands on samples from the far side of the Moon to prove their theory. "Hopefully in future, a sample return or a manned mission would certainly help to say more about which theory is more probable," Zuber said.

Although the second, smaller moon was destroyed in the collision with the larger Moon, there are asteroids that temporarily orbit Earth (Fig. 10.5).

In March of 2012, astronomers at Cornell University published the result of a computer study, suggesting that asteroids orbiting the Sun might temporarily become natural satellites of Earth. In fact, they said, Earth usually has more than one temporary moon, which they called mini-moons. These astronomers said the mini-moons would follow complicated paths around Earth for

FIG. 10.5 Artist's conception of Earth with two moons (Credit: NASA)

a time. Eventually, they would break free of Earth's gravity—only to be immediately recaptured into orbit around the Sun, becoming an asteroid once more. The little moons envisioned by these astronomers might typically be only a few feet across and might orbit our planet for less than a year before going back to orbit the Sun as asteroids.

Have astronomers detected any of these mini-moons? Yes. Writing in the magazine *Astronomy* in December 2010, Donald Yeomans (manager of NASA's Near-Earth Object Program Office at Caltech's Jet Propulsion Laboratory) described an object discovered in 2006 that appears to fit that description. The object—now designated 2006 RH120—is estimated to be 5 m (about 15 ft.) in diameter. Yeomans said that, when this object was discovered in a near-polar orbit around Earth, it was thought at first to be a third stage Saturn S-IVB booster from *Apollo 12*, but later determined to be an asteroid. 2006 RH120 began orbiting the Sun again 13 months after its discovery, but it's expected to sweep near Earth and be re-captured as a mini-moon by Earth's gravity later in this century.

Just don't confuse quasi-satellites with second moons of Earth. A quasi-satellite is an object in a co-orbital configuration with Earth (or another planet). Scientists would say there is a 1:1 orbital resonance between Earth and this object. In other words, a quasi-satellite is orbiting the Sun, just as Earth is. Its orbit around the Sun takes exactly the same time as Earth's orbit, but the shape of the orbit is slightly different.

The most famous quasi-satellite in our time—and an object you might have heard called a second moon of Earth—is 3753 Cruithne. This object is 5 km—about 3 miles—wide. Notice that it has an asteroid name. That's because it *is* an asteroid orbiting our Sun, one of several thousand asteroids whose orbits cross Earth's orbit. Astronomers discovered Cruithne in 1986, but it wasn't until 1997 that they figured out its complex orbit. It's not a second moon for Earth; it doesn't orbit Earth. But Cruithne is co-orbiting the Sun *with* Earth. Like all quasi-satellites, Cruithne orbits the Sun once for every orbit of Earth.

Earth's gravity affects Cruithne in such a way that Earth and this asteroid return every year to nearly the same place in orbit relative to each other. However, Cruithne won't collide with Earth, because its orbit is very inclined with respect to ours. It moves in and out of the plane of the ecliptic, or plane of Earth's orbit around the Sun.

Orbits like that of Cruithne aren't stable. Computer models indicate that Cruithne will spend only another 5000 years or so in its current orbit. That's a blink on the long timescale of our Solar System. The asteroid might then move into a true orbit around Earth for a time, at which time it *would* be a second Moon—but not for long. Astronomers estimate that, after 3000 years orbiting Earth, Cruithne would escape back into orbit around the Sun.

By the way, Cruithne isn't the only quasi-satellite in a 1:1 resonance orbit with Earth. The objects 2010 SO16 and (277810) 2006 FV35, among others, are also considered quasi-satellites to Earth. These objects are not second moons for Earth, although sometimes you might hear people mistakenly say they are.

Does Earth ever have more than one moon? Surprisingly, the answer is yes.

Earth usually has more than one moon. As discussed above, asteroids orbiting the Sun might temporarily become mini-moons, following complicated paths around Earth for a time. Eventually, they would break free of Earth's gravity—only to be immediately recaptured into orbit around the Sun. Astronomers simulated the orbits of miniMoons with a supercomputer and published their work in the March 2012 issue of the journal *Icarus*.

Moons are defined as Earth's *natural* satellites. They orbit Earth. The little moons envisioned by these astronomers might be only a few feet across and might orbit our planet for less than a year before going back to orbit the Sun as asteroids.

According to astronomers' simulation, most asteroids that are captured by Earth's gravity wouldn't orbit Earth in neat ellipses, as our 2000-miles (3000-km) diameter Moon does. Instead, the mini-moons, sometimes less than a meter across, would follow complicated, twisting paths.

The Moon's neat orbit comes from the fact that it's tightly held by Earth's gravity. Mini-moons would be literally tugged from many sides by Earth, the Moon, and the Sun, resulting in their complicated—and temporary—orbits. According to these astronomers, a mini-moon would remain captured by Earth until a particularly strong tug by the Sun or Moon broke the pull of Earth's gravity. At that point, the Sun would once again take control of the former mini-moon, which would then go back to being an asteroid. Although the typical mini-moon would orbit Earth for about 9 months, some of them could orbit our planet for decades, astronomers say.

Mikael Granvik (formerly at University of Hawaii Manoa and now at Helsinki), Jeremie Vaubaillon (Paris Observatory), and Robert Jedicke (UH Manoa) used a supercomputer to simulate the passage of 10 million asteroids past Earth. They said the calculations were so complex that—if you tried them on your home computer—it would take you 6 years to finish them. They concluded that at any given time there should be at least one asteroid with a diameter of at least one meter orbiting Earth. Of course, there may also be many smaller objects orbiting Earth, too.

The University of Arizona's Catalina Sky Survey discovered a mini-moon in 2006. Known to astronomers as 2006 RH120, it was about the size of a car. It orbited Earth for less than a year after its discovery, and then resumed its orbit around the Sun.

Humankind has been searching for Earth's second moon since the middle of the nineteenth century. The first major claim of another moon of Earth was made by French astronomer Frédéric Petit, director of the Toulouse Observatory, who in 1846 announced that he had discovered a second moon in an elliptical orbit around Earth. It was claimed to have also been reported by Lebon and Dassier at Toulouse, and by Larivière at Artenac Observatory during the early evening of March 21, 1846. Petit proposed that this second moon had an elliptical orbit, a period of 2 h 44 min, with 3570 km (2220 miles) apogee and 11.4 km (7.1 miles) perigee. This claim was soon dismissed by his peers. The 11.4 km perigee is similar to the cruising altitude of most modern airliners, and within Earth's atmosphere.

Petit published another paper on his 1846 observations in 1861, basing the second moon's existence on perturbations in movements of the actual Moon. This second moon hypothesis was not confirmed either. Nevertheless, Petit's proposed Moon became a major plot point in Jules Verne's 1870 science fiction novel *From the Earth to the Moon*.

In 1898 Hamburg scientist Dr. Georg Waltemath announced that he had located a system of tiny moons orbiting Earth. He had begun his search for secondary moons based on the hypothesis that something was gravitationally affecting the Moon's orbit.

Waltemath described one of the proposed moons as being 1,030,000 km (640,000 miles) from Earth, with a diameter of 700 km (430 miles), a 119-day orbital period, and a 177-day synodic period. He also said it did not reflect enough sunlight to be observed without a telescope, unless viewed at certain times, and made several predictions as to when it would appear. "Sometimes, it shines at night like the Sun but only for an hour or so." However, after the failure of a corroborating observation of Waltemath's moons by the scientific community, these objects were discredited. Especially problematic was a failed prediction that they would be observable in February 1898. The August 1898 issue of *Science* mentioned that Waltemath had sent the journal "an announcement of a third moon," which he termed a *wahrhafter Wetter und Magnet Mond* ("real weather and magnet Moon"). It was supposedly 746 km (464 miles) in diameter, and closer than the Moon that he had described previously.

In 1918, astrologer Walter Gornold, also known as Sepharial, claimed to have confirmed the existence of Waltemath's moon. He named it Lilith. Sepharial claimed that Lilith was a 'dark' moon invisible for most of the time, but he claimed to have viewed it as it crossed the Sun. In 1926 the science journal *Die Sterne* published the findings of amateur German astronomer W. Spill, who claimed to have successfully viewed a second moon orbiting Earth. In the late 1960s John Bargby claimed to have observed over ten small natural satellites of Earth, but this was not confirmed.

William Henry Pickering (1858–1938) studied the possibility of a second moon and made a general search, ruling out the possibility of many types of objects by 1903. His 1922 article "A Meteoritic Satellite" in *Popular Astronomy* resulted in increased searches for small natural satellites by amateur astronomers. Pickering had also proposed the Moon itself had broken off from Earth.

In early 1954 the U. S. Army's Office of Ordnance Research commissioned Clyde Tombaugh, discoverer of Pluto, to search for near-Earth asteroids. The army issued a public statement to explain the rationale for this survey. Donald Keyhoe, who was later director of the National Investigations Committee on Aerial Phenomena (NICAP), a UFO research group, said that his Pentagon source had told him that the actual reason for the quickly initiated search was that two near-Earth objects had been picked up on new long-range radar in mid-1953. In May 1954, Keyhoe asserted that the search had been successful, and either one or two objects had been found. At the Pentagon, a general who heard the news reportedly asked whether the satellites were natural or artificial. Tombaugh denied the alleged discovery in a letter to Willy Ley, and the October 1955 issue of *Popular Mechanics* magazine reported:

Professor Tombaugh is closemouthed about his results. He won't say whether or not any small natural satellites have been discovered. He does say, however, that newspaper reports of 18 months ago announcing the discovery of natural satellites at 400 and 600 miles out are not correct. He adds that there is no connection between the search program and the reports of so-called flying saucers.

At a 1957 meteor conference in Los Angeles, Tombaugh reiterated that his 4-year search for natural satellites had been unsuccessful. In 1959 he issued a final report stating that nothing had been found in his search. It was discovered that small bodies can be temporarily captured, as shown by 2006 RH120, which was in Earth orbit in 2006–2007. In 2010, the first known Earth trojan was discovered in data from the Wide-field Infrared Survey Explorer (WISE), and is currently called 2010 TK7.

In 2011, planetary scientists Erik Asphaug and Martin Jutzi proposed a model in which a second moon would have existed 4.5 billion years ago, and later impacted the Moon, as a part of the accretion process in the formation of the Moon.

Bottom line: That asteroid called 3753 Cruithne is not a second moon of Earth, but its orbit around the Sun is so strange that you still sometimes hear people say it is. Meanwhile, astronomers have suggested that Earth does frequently capture asteroids, which might orbit our world for about a year before breaking free of Earth's gravity and orbiting the Sun once more. Some of these asteroids might temporarily become Earth's second moon.

The Lemon-Shaped Moon

The Moon is slightly squashed, as if someone had held it at the poles between thumb and forefinger and squeezed, flattening it around its equatorial midsection. That's not surprising. The Moon spins, and the outward centrifugal force should indeed have generated a bulge as the molten magma of a young Moon cooled to solid rock eons ago.

But as far back as 1799, the mathematician Pierre-Simon Laplace noticed a back-and-forth wobbling because of the Moon's deformed shape. Although the flattening was slight—the Moon's width, 2159 miles, is about 2.5 miles greater than its pole-to-pole height—it was still greater than would be expected for its current rotation period of 27 days 7 h 43 min and 11.5 s.

"The puzzle had been the Moon was too flat," said Maria T. Zuber, a professor of geophysics and planetary sciences at the Massachusetts Institute of Technology.

Space probes of the 1960s and 1970s found a second deformity of the Moon: it is slightly elongated along the Moon-Earth axis. That is, if the Moon were sliced in half along its equator, the cross-section would not be a circle but more like a football, with one of the narrow ends pointing toward Earth (Fig. 10.6).

Fig. **10.6** The Moon is slightly squished in the middle (Credit: NASA)

However no one could come up with a completely convincing explanation for the Moon's current shape. That is not its only mystery. Another is why its near side, which always faces Earth, is so different in material and appearance from the far side. The Moon's origin still holds some uncertainties, although many scientists believe that it formed out of the debris when a Mars-size object struck Earth 4.5 billion years ago.

"Quite a lot of the darned thing is still quite mysterious," said Kimmo Innanen, a professor of astronomy at York University in Toronto.

In an issue of *Science*, Zuber, with Jack Wisdom and Ian Garrick-Bethell, say they have a possible answer to the problem of the Moon's shape. Actually, they say they have several.

What Laplace did not know is that the Moon is moving away from Earth and slowing down. Years of bouncing laser beams off mirrors left on the lunar surface by the Apollo astronauts show that each year the Moon is another 1.5 in. farther from Earth.

The Moon now orbits in what astronomers call a 1:1 resonance with Earth, its orbital period equal to its rotation time so that the same side of the Moon always faces Earth.

In the past, the Moon was much closer and took less time to orbit. With the 1:1 resonance, the Moon spun faster as well, possibly explaining the bulge. But those calculations did not come with the correct answer for the observed distortion along the Moon-Earth axis. "They ought to be self-consistent," Dr. Zuber said.

One suggestion has been that the Moon by chance cooled into this somewhat odd configuration. Other solid planets, such as Earth, are not exactly in their predicted shape either.

The M.I.T. scientists, however, say the observed distortions are larger than would be expected for chance. Instead, they suggest that in the Moon's early history, it traveled in an elliptical rather than a circular orbit and that it was in a 3:2 resonance, spinning three times for every two orbits. It would have been in this resonance for only a few hundred million years at most, the scientists said, before tidal forces slowed its spin and it fell into the current 1:1 circular resonance. Their calculations show that such an orbit would provide the necessary forces to produce the Moon's shape. "There's a 200-year-old problem, and we've got the first solution that works," Dr. Zuber said. They also found that orbits with higher resonances, like two spins for every orbit, could produce the same lunar shape. "It's a bunch of families of solutions," Dr. Zuber said.

In an accompanying commentary in *Science*, Dr. Innanen described the proposed solution as "ingenious." But Peter M. Goldreich, an emeritus professor of astrophysics and planetary physics at the California Institute of Technology, said the M.I.T. team did not explain how the Moon was caught in the 3:2 resonance. "That is a real weakness," he said.

Although an elliptical orbit and a 3:2 resonance may explain the shape, they could not explain the disparities between the Moon's sides. The crust on the near side is much thinner, with vast plains of dark basalt, called maria, that erupted long ago. The crust of the far side is thicker, with many more craters and fewer maria.

Partly because of the denser basalt of the mare, the Moon's center of mass is not at the center, but shifted by more than a mile. But how that shift occurred is unknown.

"That would be a good research project," Dr. Innanen said.

Many of the mysteries depend on a better understanding of the Moon's early history. Under the prevailing theory that the Moon formed after something the size of Mars slammed into Earth, scientists believe that the Moon initially orbited very close, perhaps just 16,000 miles from Earth (compared with roughly 238,000 miles today) before moving outward.

David J. Stevenson, a professor of planetary science at Caltech, said the evidence for the collision theory was solid. Then he stopped. He said instead, "The argument against the alternatives is strong."

Other studies have also confirmed the lopsided shape of the Moon. Scientists combined observations from two NASA missions to check out the Moon's lopsided shape and how it changes under Earth's sway—a response not seen from orbit before.

The team drew on studies by NASA's Lunar Reconnaissance Orbiter (LRO), which has been investigating the Moon since 2009, and by NASA's Gravity Recovery and Interior Laboratory (GRAIL) mission. Because orbiting spacecraft gathered the data, the scientists were able to take the entire Moon into account, not just the side that can be observed from Earth.

"The deformation of the Moon due to Earth's pull is very challenging to measure, but learning more about it gives us clues about the interior of the Moon," said Erwan Mazarico from the Massachusetts Institute of Technology in Cambridge. The lopsided shape of the Moon is one result of its gravitational tug-of-war with Earth. The mutual pulling of the two bodies is powerful enough to stretch them both, so they wind up shaped a little like two eggs with their ends pointing toward one another. On Earth, the tension has an especially strong effect on the oceans, because water moves so freely; this tug-of-war drives Earth's tides.

Earth's distorting effect on the Moon, called the lunar body tide, is more difficult to detect because the Moon is solid except for its small core. Even so, there is enough force to raise a bulge about 20 in. (51 cm) high on the nearside of the Moon and create a similar one on the far side.

The position of the bulge actually shifts a few inches over time. Although the same side of the Moon constantly faces Earth, because of the tilt and shape of the Moon's orbit, the side facing Earth appears to wobble. From the Moon's viewpoint, our planet doesn't sit motionless but instead moves around within a small patch of sky. The bulge responds to Earth's movements like a dance partner, following wherever the leader goes.

"If nothing changed on the Moon—if there were no lunar body tide or if its tide were completely static—then every time scientists measured the surface height at a particular location, they would get the same value," said Mike Barker from Sigma Space Corporation based at NASA's Goddard Space Flight Center in Greenbelt, MD.

A few studies of these subtle changes were conducted previously from Earth, but not until LRO and GRAIL did satellites provide enough resolution to see the lunar tide from orbit.

To search for the tide's signature, the scientists turned to data taken by LRO's Lunar Orbiter Laser Altimeter (LOLA), which mapped the height of features on the Moon's surface. The team chose spots that the spacecraft had passed over more than once, each time approaching along a different flight path. More than 350,000 locations were selected, covering areas on the near side and far side of the Moon.

The researchers precisely matched measurements taken at the same spot and calculated whether the height had risen or fallen from one satellite pass to the next; a change indicated a shift in the location of the bulge.

A crucial step in the process was to pinpoint exactly how far above the surface LRO was located for each measurement. To reconstruct the spacecraft's orbit with sufficient accuracy, the researchers needed the detailed map of the Moon's gravity field provided by the GRAIL mission.

"This study provides a more direct measurement of the lunar body tide and much more comprehensive coverage than has been achieved before," said John Keller from Goddard.

The good news for lunar scientists is that the new results are consistent with earlier findings. The estimated size of the tide confirmed the previous measurement of the bulge. The other value of great interest to researchers is the overall stiffness of the Moon, known as the "Love number," h_2, and this was also similar to prior results.

Having confirmation of the previous values, with significantly smaller measurement uncertainties than before, will make the lunar body tide a more useful piece of information for scientists.

"This research shows the power of bringing together the capabilities of two missions. The extraction of the tide from the LOLA data would have been impossible without the gravity model of the Moon provided by the GRAIL mission," said David Smith, who is affiliated with Goddard and the Massachusetts Institute of Technology.

Why Is the Moon Two-Faced?

The Man in the Moon was born when cosmic impacts struck the near side of the Moon, the side that faces Earth. These collisions punched holes in the Moon's crust, which later filled with vast lakes of lava that formed the dark areas known as maria, or seas. In 1959, when the Soviet spacecraft *Luna 3* transmitted the first images of the "dark," or far side, of the Moon, the side facing away from Earth, scientists immediately noticed fewer maria there. This mystery—why no Man in the Moon exists on the Moon's far

FIG. 10.7 Mare Orientale is a vast impact basin, the hole left by an asteroid that hit the Moon about 3 billion years ago. Looking like a humongous bulls-eye, it's a multi-ring crater, and the outer ramparts are a full 950 km (590 miles) across (Credit: NASA)

side—is called the Lunar Far side Highlands Problem. For decades, scientists have been trying to understand why the Earth-facing side of the Moon looks so different from the Moon's far side. Now, they may have an answer to that mystery (Fig. 10.7).

"I remember the first time I saw a globe of the Moon as a boy, being struck by how different the far side looks," said study co-author Jason Wright, an astronomer at Pennsylvania State University. "It was all mountains and craters. Where were the maria?"

Now scientists may have solved the 55-year-old mystery—heat from the young Earth as the newborn Moon was cooling caused the difference. The researchers came up with the solution during their work on exoplanets, which are worlds outside the Solar System.

"There are many exoplanets that are really close to their host stars," said lead study author Arpita Roy, also of Penn State. "That really affects the geology of those planets."

Similarly, the Moon and Earth are generally thought to have orbited very close together after they formed. The leading idea explaining the Moon's formation suggests that it arose shortly after the nascent Earth collided with a Mars-size planet about 4.5 billion years ago, with the resulting debris coalescing into the Moon. Scientists say the newborn Moon and Earth were 10–20 times closer to each other than they are now.

"The Moon and Earth loomed large in each other's skies when they formed," Roy said. Since the Moon was so close to Earth, the mutual pull of gravity was strong. The gravitational tidal forces the Moon and Earth exerted on each other braked their rotations, resulting in the Moon always showing the same face to Earth, a situation known as tidal lock.

The Moon and Earth were very hot shortly after the giant impact that formed the Moon. The Moon, being much smaller than Earth, cooled more quickly. Since the Moon and Earth were tidally locked early on, the still-hot Earth—more than 4530 °F (2500 °C)—would have cooked the near side of the Moon, keeping it molten. On the other hand, the far side of the Moon would have cooled, albeit slowly.

The difference in temperature between the Moon's halves influenced the formation of its crust. The lunar crust possesses high concentrations of aluminum and calcium, elements that are very hard to vaporize.

"When rock vapor starts to cool, the very first elements that snow out are aluminum and calcium," said study co-author Steinn Sigurdsson of Penn State.

Aluminum and calcium would have more easily condensed in the atmosphere on the colder far side of the Moon. Eventually, these elements combined with silicates in the mantle of the Moon to form minerals known as plagioclase feldspars, making the crust of the far side about twice as thick as that of the near side. "The heat of Earth soon after the giant impact was a really important factor shaping the Moon," Roy said.

When collisions from asteroids or comets blasted the Moon's surface, they could punch through the near side's crust to generate maria. In contrast, impacts on the far side's thicker crust failed to penetrate deeply enough to cause lava to well up, instead leaving the far side of the Moon with a surface of valleys, craters, and highlands, but almost no maria.

"It's really cool that our understanding of exoplanets is affecting our understanding of the Solar System," Roy said. Future research could generate detailed 3D models testing this idea, Roy suggested. The authors detailed their findings in the *Astrophysical Journal Letters*.

The far side's terrain is rugged, with a multitude of impact craters and relatively few flat lunar maria. It has one of the largest craters in the Solar System, the South Pole-Aitken Basin.

About 18 % of the far side is occasionally visible from Earth due to libration. The remaining 82 % remained unobserved until 1959, when the Soviet Union's *Luna 3* space probe photographed it. The Russian Academy of Sciences published the first atlas of the far side in 1960. In 1968, the *Apollo 8* mission's astronauts were the first humans to view this region directly when they orbited the Moon. To date, no one has explored the far side of the Moon on the ground. Astronomers have suggested installing a large radio telescope on the far side, where the Moon would shield it from possible radio interference from Earth.

In total 59 % of the Moon's surface is visible from Earth at one time or another. However, useful observation of the parts of the far side of the Moon occasionally visible from Earth is difficult because of the low viewing angle from Earth (they cannot be observed "full on").

The commonly used term "dark side of the Moon" does not refer to "dark" as in the absence of light, but rather as the unknown, as, until we were able to send spacecraft around the Moon this area had never been seen. Although many misconstrue this to think that the "dark side" receives little to no sunlight, in reality, both the near and far sides receive (on average) almost equal amounts of light directly from the Sun. However, the near side also receives sunlight reflected from Earth, known as earthshine. Earthshine does not reach the area of the far side, which cannot be seen from Earth. Only during a full Moon (as viewed from Earth) is the whole far side of the Moon physically dark. The word "dark" may also refer to the fact that communication with spacecraft can be blocked while on the far side of the Moon, as it was during Apollo space missions.

The two hemispheres have distinctly different appearances, with the near side covered in multiple, large maria. The far side has a battered, densely cratered appearance with few maria. Only 1 % of the surface of the far side is covered by maria, compared to 31.2 % on the near side. One commonly accepted explanation for this difference is related to a higher concentration of heat-producing elements on the near-side hemisphere, as has been demonstrated by geochemical maps obtained from the Lunar Prospector gamma-ray spectrometer. Although other factors such as surface elevation and crustal thickness could also affect where basalts erupt, these do not explain why the far side South Pole-Aitken Basin (which contains the lowest elevations of the Moon and possesses a thin crust) was not as volcanically active as Oceanus Procellarum on the near side.

It has also been proposed that the differences between the two hemispheres may have been caused by a collision with a smaller companion that also originated from the Theia collision. In this model the impact led to an accretionary pile rather than a crater, contributing a hemispheric layer of extent and thickness that may be consistent with the dimensions of the far side highlands.

The far side has more visible craters. This was thought to be a result of the effects of lunar lava flows, which cover and obscure craters, rather than a shielding effect from Earth. NASA calculates that Earth obscures only about 4 square degrees out of 41,000 square degrees of the sky as seen from the Moon. "This makes Earth negligible as a shield for the Moon. It was suggested that each side of the Moon has received equal numbers of impacts, but the resurfacing by lava resulted in fewer craters visible on the near side, even though both sides have received the same number of impacts." Newer research shows that the reason is heat from Earth at the time when the Moon was created.

Until the late 1950s, little was known about the far side of the Moon. Librations of the Moon periodically allowed limited glimpses of features near the lunar limb on the far side. These features, however, were seen from a low angle, hindering useful observation. (It proved difficult to distinguish a crater from a mountain range.) The remaining 82 % of the surface on the far side remained unknown, and its properties were subject to much speculation.

An example of a far side feature that can be seen through libration is the Mare Orientale, which is a prominent impact basin spanning almost 1000 km (600 miles), yet this was not even named as a feature until 1906, by Julius Franz in *Der Mond*. The true nature of the basin was discovered in the 1960s when rectified images were projected onto a globe. It was photographed in fine detail by Lunar Orbiter 4 in 1967.

On October 7, 1959, the Soviet probe *Luna 3* took the first photographs of the lunar far side, 18 of them resolvable, covering one-third of the surface invisible from the Earth. The images were analyzed, and the first atlas of the far side of the Moon was published by the USSR Academy of Sciences on November 6, 1960. It included a catalog of 500 distinguished features of the landscape. A year later the first globe containing lunar features invisible from Earth was released in the USSR, based on images from *Luna 3*.

On July 20, 1965, another Soviet probe, *Zond 3*, transmitted 25 pictures of very good quality of the lunar far side, with much better resolution than those from *Luna 3*. In particular, they revealed chains of craters, hundreds of km in length. In 1967 the second part of the "Atlas of the Far Side of the Moon" was published in Moscow, based on data from *Zond 3*, with the catalog now including 4000 newly discovered features of the lunar far side landscape. In the same year the first "Complete Map of the Moon" (1:5,000,000 scale) and updated complete globe (1:10,000,000 scale), featuring 95 % of the lunar surface, were released in the Soviet Union.

As many prominent landscape features of the far side were discovered by Soviet space probes, Soviet scientists selected names for them. This caused some controversy, and the International Astronomical Union, leaving many of those names intact, later assumed the role of naming lunar features on this hemisphere.

The far side was first seen directly by human eyes during the Apollo 8 mission in 1968. Astronaut William Anders described the view: "The backside looks like a sand pile my kids have played in for some time. It's all beat up, no definition, just a lot of bumps and holes."

Since that time it has been seen by all crew members of the *Apollo 8* and *Apollo 10* through *Apollo 17* missions, and photographed by multiple lunar probes. Spacecraft passing behind the Moon were out of direct radio communication with Earth and had to wait until the orbit allowed transmission.

During the Apollo missions, the main engine of the service module was fired when the vessel was behind the Moon, producing some tense moments in Mission Control before the craft reappeared.

Geologist-astronaut Harrison Schmitt, who became the second to last to step onto the Moon, had aggressively lobbied for his landing site to be on the far side of the Moon, targeting the lava-filled crater Tsiolkovsky. Schmitt's ambitious proposal included a special communications satellite based on the existing TIROS satellites to be launched into an orbit around the L2 Lagrange point so as to maintain line-of-sight contact with the astronauts during their powered descent and lunar surface operations. NASA administrators rejected these plans based on added risk and lack of funding.

Because the far side of the Moon is shielded from radio transmissions from Earth, it is considered a good location for placing radio telescopes for use by astronomers. Small, bowl-shaped craters provide a natural formation for a stationary telescope similar to Arecibo in Puerto Rico. For much larger-scale telescopes, the 100-km (62 miles) diameter crater Daedalus is situated near the center of the far side, and the 3-km (2 miles)-high rim would help to block stray communications from orbiting satellites. Another potential candidate for a radio telescope is the Saha Crater.

Before deploying radio telescopes to the far side, several problems must be overcome. The fine lunar dust can contaminate equipment, vehicles, and spacesuits. The conducting materials used for the radio dishes must also be carefully shielded against the effects of solar flares. Finally the area around the telescopes must be protected against contamination by other radio sources.

The L_2 Lagrange point of the Earth-Moon system is located about 62,800 km (39,000 miles) above the far side, which has also been proposed as a location for a future radio telescope which would perform an orbit about the Lagrangian point.

Down the Lunar Rabbit Hole

Alice discovered a whole new world when she followed the White Rabbit down the hole. It makes you wonder. What's waiting down the rabbit hole on the Moon? NASA's Lunar Reconnaissance Orbiter (LRO) is beaming back images of caverns hundreds of feet deep—beckoning scientists to follow.

"They could be entrances to a geologic wonderland," says Mark Robinson of Arizona State University, principal investigator for the LRO camera. "We believe the giant holes are skylights that formed when the ceilings of underground lava tubes collapsed" (Fig. 10.8).

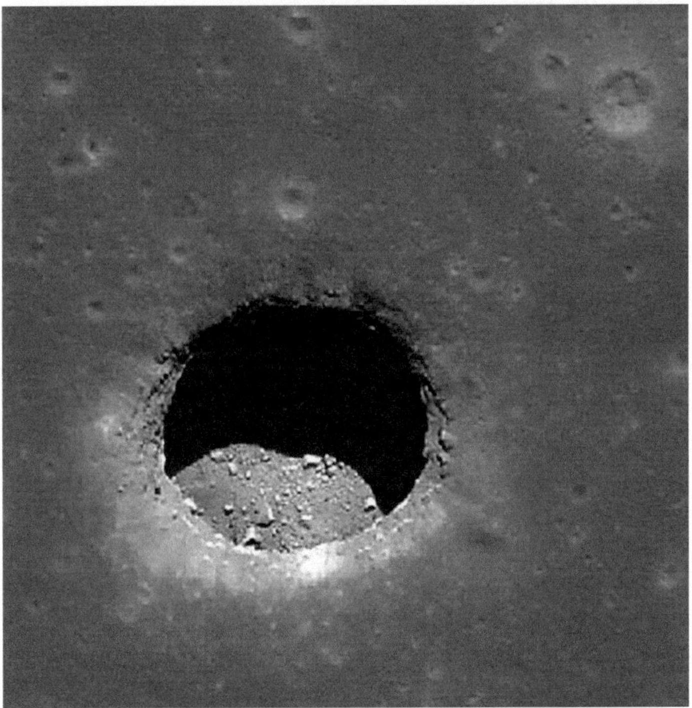

Fig. 10.8 Boulders litter the floor of a 300-foot-deep lunar pit crater in this image from the Lunar Reconnaissance Orbiter, released on September 14, 2010. The pit is located in the Mare Tranquilatis area of the Moon and was most likely created by the collapse of a subsurface lava tube (Credit: NASA)

Japan's Kaguya spacecraft first photographed the enormous caverns. Now the powerful Lunar Reconnaissance Orbiter Camera (LROC, the same camera that photographed Apollo landers and astronauts' tracks in the moondust) is giving us enticing high-resolution images of the caverns' entrances and their surroundings.

Back in the 1960s, before humans set foot on the Moon, researchers proposed the existence of a network of tunnels, relics of molten lava rivers beneath the lunar surface. They based their theory on early orbital photographs that revealed hundreds of long, narrow channels called rilles winding across the vast lunar plains, or maria. Scientists believed these rilles to be surface evidence of below-ground tunnels through which lava flowed billions of years ago.

"It's exciting that we've now confirmed this idea," says Robinson. "The Kaguya and LROC photos prove that these caverns are skylights to lava tubes, so we know such tunnels can exist intact at least in small segments after several billion years."

Lava tubes are formed when the upper layer of lava flowing from a volcano starts to cool while the lava underneath continues to flow in tubular channels. The hardened lava above insulates the molten lava below, allowing it to retain its liquid warmth and continue flowing. Lava tubes are found on Earth and can vary from a simple tube to a complex labyrinth that extends for miles.

If the tunnels leading off the skylights have stood the test of time and are still open, they could someday provide human visitors protection from incoming meteoroids and other perils.

"The tunnels offer a perfect radiation shield and a very benign thermal environment," says Robinson. "Once you get down to 2 m under the surface of the Moon, the temperature remains fairly constant, probably around –30 to –40 °C."

That may sound cold, but it would be welcome news to explorers seeking to escape the temperature extremes of the lunar surface. At the Moon's equator, mid-day temperatures soar to 100 °C and plunge to a frigid –150 °C at night.

Paul Spudis of the Lunar and Planetary Institute agrees that lunar lava tubes and chambers hold potential advantages to future explorers but says, "Hold off on booking your next vacation at the Lunar Carlsbad Hilton. Many tunnels may have filled up with their own solidified lava."

However, like Alice's White Queen, who "believed as many as six impossible things before breakfast," Spudis is keeping an open mind.

"We just can't tell, with our remote instruments, what the skylights lead to. To find out for sure, we'd need to go to the Moon and do some spelunking. I've had my share of surprises in caving. Several years ago I was helping map a lava flow in Hawaii. We had a nice set of vents, sort of like these skylights. It turned out that there was a whole new cave system that was not evident from aerial photos."

As for something similar under the lunar skylights? "Who knows?" says Spudis. "The Moon continually surprises me."

This could be a white rabbit worth following.

The mysterious pits range in size from about 5 m (~5 yd.) across to more than 900 m (~984 yd.) in diameter, and three of them were first identified using images from the Kaguya spacecraft. Hundreds more were found using a new computer algorithm that automatically scanned thousands of high-resolution images of the lunar surface from LRO's Narrow Angle Camera (NAC).

"Pits would be useful in a support role for human activity on the lunar surface," said Robert Wagner of Arizona State University. "A habitat placed in a pit—ideally several dozen m back under an overhang—would provide a very safe location for astronauts: no radiation, no micrometeorites, possibly very little dust, and no wild day-night temperature swings." Wagner developed the computer algorithm and is lead author of a paper on this research available online in the journal *Icarus*.

Most pits were found either in large craters with impact melt ponds—areas of lava that formed from the heat of the impact and later solidified, or in the lunar maria—dark areas on the moon that are extensive solidified lava flows hundreds of miles across. Various cultures have interpreted the patterns formed by the maria features in different ways; for example, some saw the face of a man, while others saw a rabbit or a boy carrying a bundle of sticks on his back.

The pits could form when the roof of a void or cave collapses, perhaps from the vibrations generated by a nearby meteorite impact, according to Wagner. However, he noted that from their appearance in the LRO photos alone, there is little evidence to point to any particular cause. The voids could have been created when molten rock flowed under the lunar surface; on Earth, lava tubes form when magma flows beneath a solidified crust and later drains away. The same process could happen on the Moon, especially in a large impact crater, the interior of which can take hundreds of thousands of years to cool, according to Wagner. After an impact crater forms, the sides slump under lunar gravity, pushing up the crater's floor and perhaps causing magma to flow under the surface, forming voids in places where it drains away.

Exploring impact melt pits would pin down the nature of the voids in which they form. "They are likely due to melt flow within the pond from uplift after the surface has solidified, but before the interior has cooled," said Wagner. "Exploring impact melt pits would help determine the magnitude of this uplift, and the amount of melt flow after the pond is in place."

Exploring the pits could also reveal how oceans of lava formed the lunar maria. "The mare pits in particular would be very useful for understanding how the lunar maria formed. We've taken images from orbit looking at the walls of these pits, which show that they cut through dozens of layers, confirming that the maria formed from lots of thin flows, rather than a few big ones. Ground-level exploration could determine the ages of these layers, and might even find solar wind particles that were trapped in the lunar surface billions of years ago," said Wagner.

To date, the team has found over 200 pits spread across the melt ponds of 29 craters, all of which are considered geologically young "Copernican" craters at less than a billion years old; eight pits in the lunar maria, three of which were previously known from images from the Kaguya orbiter; and two pits in highlands terrain.

The researchers found most pits in either large craters with impact melt ponds, which are areas of lava that formed from the heat of the impact and later solidified, or in the lunar maria.

The general age sequence matches well with the pit distributions, according to Wagner. "Impact melt ponds of Copernican craters are some of the younger terrains on the Moon, and while the maria are much older at around 3 billion years old, they are still younger and less battered than the highlands. It's possible that there's a 'sweet spot' age for pits, where enough impacts have occurred to create a lot of pits, but not enough to destroy them," said Wagner.

There are almost certainly more pits out there, given that LRO has only imaged about 40 % of the Moon with appropriate lighting for the automated pit searching program, according to Wagner. He expects there may be at least two to three more mare pits and several dozen to over a hundred more impact melt pits, not including any pits that likely exist in already-imaged areas, but are too small to conclusively identify even with the NAC's resolution.

"We've taken images from orbit looking at the walls of these pits, which show that they cut through dozens of layers, confirming that the maria formed from lots of thin flows, rather than a few big ones. Ground-level exploration could determine the ages of these layers, and might even find solar wind particles that were trapped in the lunar surface billions of years ago," said Wagner. The exploration of maria pits could also reveal how the lunar maria was formed, the researcher noted.

"We'll continue scanning NAC images for pits as they come down from the spacecraft, but for about 25 % of the Moon's surface area (near the poles) the Sun never rises high enough for our algorithm to work," said Wagner. "These areas will require an improved search algorithm, and even that may not work at very high latitudes, where even a human has trouble telling a pit from an impact crater."

The next step would be to tie together more datasets such as composition maps, thermal measurements, gravity measurements, etc., to gain a better understanding of the environments in which these pits form, both at and below the surface, according to Wagner.

"The ideal follow-up, of course, would be to drop probes into one or two of these pits, and get a really good look at what's down there," adds Wagner. "Pits, by their nature, cannot be explored very well from orbit—the lower walls and any floor-level caves simply cannot be seen from a good angle. Even a few pictures from ground-level would answer a lot of the outstanding questions about the nature of the voids that the pits collapsed into. We're currently in the very early design phases of a mission concept to do exactly this, exploring one of the largest mare pits."

NASA has awarded a contract to Astrobotic to develop technologies for exploring caves on the Moon, Mars, and beyond. Astrobotic will detail a mission concept for entering a planetary cave through a skylight, and exploring and modeling the interior.

"Skylights are gateways to wonders of exploration, science, and resources that await beneath planetary surfaces," said Red Whittaker, Astrobotic CEO. "Robots are our access to those new worlds." Robotic technologies will be developed to explore the extreme terrains of skylights and caves. This is very different from surface exploration, as has been achieved on the Moon and Mars. Technologies will be developed to descend into the holes, negotiate the blocky floors, and thread into the tunnels. The company will also roadmap technology for future planetary cave exploration missions. Astrobotic will collaborate with experts in subterranean robotics at Carnegie Mellon University on this contract.

New Origin for the Ocean of Storms

Using data from NASA's Gravity Recovery and Interior Laboratory (GRAIL) mission scientists have solved a lunar mystery almost as old as the Moon itself.

Early theories suggested the craggy outline of a region of the Moon's surface known as Oceanus Procellarum, or the Ocean of Storms, was caused by an asteroid impact. If this theory had been correct, the basin it formed would be the largest asteroid impact basin on the Moon. However, mission scientists studying GRAIL data believe they have found evidence the craggy outline of this rectangular region—roughly 1600 miles (2600 km) across—is actually the result of the formation of ancient rift valleys.

"The nearside of the Moon has been studied for centuries, and yet continues to offer up surprises for scientists with the right tools," said Maria Zuber, principal investigator of NASA's GRAIL mission. "We interpret the gravity anomalies discovered by GRAIL as part of the lunar magma plumbing system—the conduits that fed lava to the surface during ancient volcanic eruptions."

The surface of the Moon's nearside is dominated by a unique area called the Procellarum region, characterized by low elevations, unique composition, and numerous ancient volcanic plains. It is the only one of the lunar maria to be called an ocean. That's because it's the largest of the maria.

The rifts are buried beneath dark volcanic plains on the near-side of the Moon and have been detected only in the gravity data provided by GRAIL. The lava-flooded rift valleys are unlike anything found anywhere else on the Moon and may at one time have resembled rift zones on Earth, Mars and Venus. The findings are published online in the journal *Nature*.

Another theory arising from recent data analysis suggests this region formed as a result of churning deep in the interior of the Moon that led to a high concentration of heat-producing radioactive elements in the crust and mantle of this region. Scientists studied the gradients in gravity data from GRAIL, which revealed a rectangular shape in resulting gravitational anomalies.

"The rectangular pattern of gravity anomalies was completely unexpected," said Jeff Andrews-Hanna, a GRAIL co-investigator at the Colorado School of Mines in Golden, Colorado, and lead author of the paper. "Using the gradients in the gravity data to reveal the rectangular pattern of anomalies, we can now clearly and completely see structures that were only hinted at by surface observations."

The rectangular pattern, with its angular corners and straight sides, contradicts the theory that Procellarum is an ancient impact basin, since such an impact would create a circular basin. Instead, the new research suggests that processes beneath the Moon's surface dominated the evolution of this region. Over time, the region would cool and contract, pulling away from its surroundings and creating fractures similar to the cracks that form in mud as it dries out, but on a much larger scale.

The study also noted a surprising similarity between the rectangular pattern of structures on the Moon and those surrounding the south polar region of Saturn's icy moon Enceladus. Both patterns appear to be related to volcanic and tectonic processes operating on their respective worlds.

"Our gravity data are opening up a new chapter of lunar history, during which the Moon was a more dynamic place than suggested by the cratered landscape that is visible to the naked eye," said Andrews-Hanna. "More work is needed to understand the cause of this newfound pattern of gravity anomalies, and the implications for the history of the Moon."

Launched as GRAIL A and GRAIL B in September 2011, the probes, renamed Ebb and Flow, operated in a nearly circular orbit near the poles of the Moon at an altitude of about 34 miles (55 km) until their mission ended in December 2012. The distance between the twin probes changed slightly as they flew over areas of greater and lesser gravity caused by visible features, such as mountains and craters, and by masses hidden beneath the lunar surface.

The twin spacecraft flew in a nearly circular orbit until the end of the mission on December 17, 2012, when the probes intentionally were sent into the Moon's surface. NASA later named the impact site in honor of late astronaut Sally K. Ride, who was America's first woman in space and a member of the GRAIL mission team.

GRAIL's prime and extended science missions generated the highest resolution gravity field map of any celestial body. The map will provide a better understanding of how Earth and other rocky planets in the Solar System formed and evolved.

The GRAIL mission was managed by the Jet Propulsion Laboratory (JPL) in Pasadena, California, for NASA's Science Mission Directorate in Washington. The mission was part of the Discovery program managed at NASA's Marshall Space Flight Center in Huntsville, AL. GRAIL was built by Lockheed Martin Space Systems in Denver.

Moon Bumps

Earth's gravitational pull is so powerful that it creates a small bulge on the surface of the Moon. For the first time, scientists have observed this bump from orbit, using NASA satellites (Fig. 10.9).

The gravitational tug-of-war between Earth and the Moon is enough to stretch both celestial bodies, so they each end up having a slight oval shape, with the tapered ends facing each other.

On Earth, this gravitational tension shows up in the form of tides. The Moon's pull has a strong effect on Earth's oceans because water has so much freedom of movement. The corresponding distorting effect on the Moon, called the lunar body tide, is more difficult to see, because the Moon is solid except for a molten core. But Earth's pull raises a small bulge about 20 in. (50 cm) from the

Fig. 10.9 Earth's gravitational pull is so powerful that it creates a small bulge on the surface of the Moon, as shown in this photo. For the first time, scientists have observed this bump from orbit, using NASA satellites (Credit: NASA)

surface on the near side of the Moon and a matching bulge on the far side.

"The deformation of the Moon due to Earth's pull is very challenging to measure, but learning more about it gives us clues about the interior of the Moon," Erwan Mazarico, a scientist who works at NASA's Goddard Space Flight Center, said in a statement.

The same side of the Moon always faces Earth, but the bulge does move around a few inches over time, wobbling and following Earth's pull like a magnet, as the Moon shifts slightly during its orbit.

Scientists observed the lunar body tide using NASA's Lunar Reconnaissance Orbiter satellite, which is mapping the height of features on the Moon's surface, and NASA's Gravity Recovery and Interior Laboratory satellite, which is mapping the Moon's gravitational field. The satellites measured the height of 350,000 locations spread across areas of the Moon closest to Earth and areas of the Moon on the opposite side from Earth. The satellites passed over each location several times, so scientists could compare the height of each spot from one satellite pass to the next. By identifying which spots changed height, the researchers could track the lunar tide.

Scientists have studied the lunar body tide phenomenon from Earth, but this is the first time satellites have captured images of the lunar tide from orbit. The findings confirm the calculations scientists had already made from ground observations.

A Colorful World

Why is the Moon so many different colors? If you're looking at it during the daylight, the Moon will look faint and white surrounded by the blue of the sky. If it's night, the Moon will look bright yellow. Why does the color of the Moon seem to change from white to yellow when you go from day to night? And why does the Moon look gray in many photographs, especially the ones from space? What color is the Moon? (Fig. 10.10)

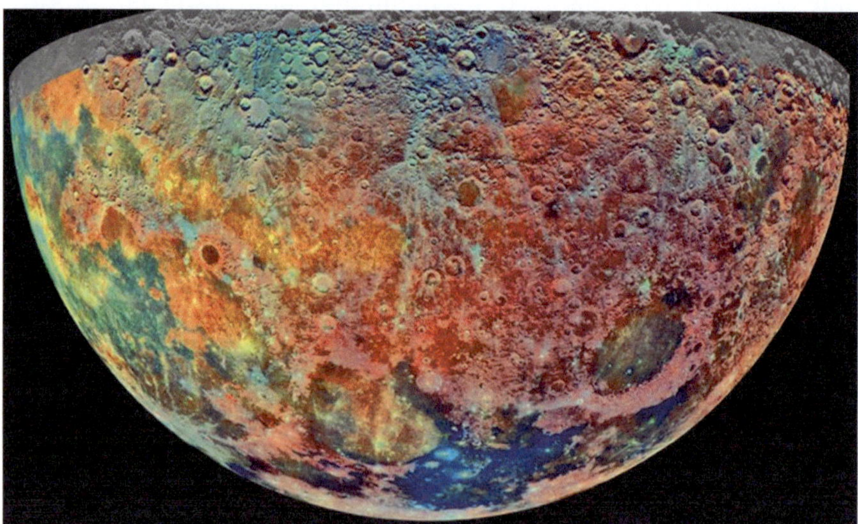

FIG. 10.10 As the exaggerated false-color composite image shows, slight differences in lunar color can be used to map the composition of lunar rocks from a distance. The image is a view of the lunar near side taken by the Galileo spacecraft as it passed above the north pole of the Moon on its way to Jupiter in 1992. *Blue* and *orange* shades indicate volcanic lava flows. The *dark blue* of Mare Tranquillitatis, at bottom center, indicates that it is richer in titanium-bearing minerals than the *green* and *orange* maria to its left. Above and to the right of Mare Tranquillitatis, the *dark oval* of Mare Crisium is surrounded by *pink* colors indicative of the iron-poor, aluminum-rich feldspars that make up the lunar highlands

The photographs of the Moon taken from space are the best true-color views of the Moon. That gray color you see comes from the surface of the Moon, which is mostly oxygen, silicon, magnesium, iron, calcium, and aluminum. The lighter color rocks are usually plagioclase feldspar, while the darker rocks are pyroxene. Most of the rocks that you can see are volcanic, and were extruded from the inside of the Moon during volcanic eruptions. Some rare rocks called olivine are actually green.

The dark regions you see on the Moon, the lunar maria, were formed by ancient volcanic eruptions. They're less reflective than the lunar highlands, and so they appear darker to the eye. The maria cover about 16 % of the lunar surface, mostly on the side we can see from Earth. Astronomers think the lunar maria were formed about 3–3.5 billion years ago, when the Moon was much more volcanically active.

When you see the Moon from here on Earth, the atmosphere partially blocks your view.

The particles in the atmosphere scatter certain wavelengths of light, and permit other wavelengths to get through directly. When the Moon is low in the sky, you're seeing its light go through the most atmosphere. Light on the blue end of the spectrum is scattered away, while the red light isn't scattered. This is why the Moon looks redder. As it goes higher in the sky, the Moon is obscured by less and less atmosphere, so it turns yellower. The same thing happens to the Sun as it rises in the sky.

At times the Moon might look red, orange, or even blue. The Moon appears to be certain colors during certain times of the year. For instance, the Harvest Moon during the fall appears very large and orange. There are two reasons for this—the Moon's path across the sky, and the climate of Earth. During certain times of the year, the Moon will rise and set at different angles. Sometimes the Moon stays really low in the sky and never reaches an overhead position. Earth's atmosphere also goes through certain changes at certain times of the year.

In some months, the atmosphere has more dust particles than usual; in other months, the atmosphere contains a lot more cloud particles than usual. Extra particles in the atmosphere mean more scattering of light. In the fall, many farmers are harvesting their crops. Lots of dust from the soil of the crops gets disturbed.

The dust floats into the atmosphere. At the same time, the Moon is lower in the sky during the fall season. So if there's more dust in the sky and the Moon is closer to the horizon, then what color will the Moon be? Orange! That's where the Harvest Moon gets its name.

You might not have heard the term before, but there astronomers sometimes use the term "black Moon." Here are some of the situations that might call for this.

When you get the second new Moon in a month, some astronomers call that a black Moon. This is similar to when you get a blue Moon, for the second full Moon in a month. The Moon takes only 29 days to go from full Moon to new Moon and back again. This is shorter than an average month (30 days or so). When you get a new Moon in the first couple of days in a month, then you can get another near the end of the month. It's that second occurrence that would be considered a black Moon. Like blue Moons, they happen about once every 2.5 years.

Another use for the term black Moon is when you get a month that doesn't have a full Moon. This can happen in February, because it has 28 days, on average, and the Moon takes 29 days to go through its phases. These occurrences are very rare. The last one happened in February 1999, and the next will occur in February 2018.

Most of the time, the Moon is a bright yellow color because it is reflecting light from the Sun. But sometimes the Moon can turn a beautiful dramatic red color. What's going on? What causes a red Moon?

There are few different situations that can cause a red Moon. The most common way to see the Moon turn red is when it is low in the sky, just after moonrise or before it's about to set below the horizon. Just like the Sun, light from the Moon has to pass through a greater mass of atmosphere when it's down near the horizon, compared to when it's overhead. Earth's atmosphere can scatter sunlight, and since moonlight is just scattered sunlight, it can scatter that, too. Red light can pass through the atmosphere and not get scattered much, while light at the blue end of the spectrum is more easily scattered. When you see a red Moon, you're seeing the red light that wasn't scattered, but the blue and green light have been scattered away. That's why the Moon looks red.

Moon Still Hot Inside

An international research team, led by Dr. Yuji Harada from Planetary Science Institute, China University of Geosciences, has found that there is an extremely soft layer deep inside the Moon and that heat is effectively generated in that layer by the gravity of Earth. These results were derived by comparing the deformation of the Moon as precisely measured by Kaguya and other probes with theoretically calculated estimates. These findings suggest that the interior of the Moon has not yet cooled and hardened, and also that it is still being warmed by the effect of Earth on the Moon. This research provides a chance to reconsider how both Earth and the Moon have been evolving since their births through mutual influence until now.

When it comes to clarifying how a celestial body like a planet or a natural satellite is born and grows, it is necessary to know as precisely as possible its internal structure and thermal state. How can we know the internal structure of a celestial body far away from us? We can get clues about its internal structure and state by thoroughly investigating how its shape changes due to external forces.

The gravitational force of a celestial body that changes the shape of another body is called a tide. For example, the ocean tide on Earth is one tidal phenomenon caused by the gravitational force between the Moon, the Sun, and Earth. Sea water is so deformable that its displacement can be easily observed. How much a celestial body can be deformed by tidal force, in this way, depends on its internal structure, and especially on the hardness of its interior. Conversely, it means that observing the degree of deformation enables us to learn about the interior, which is normally not directly visible to the naked eye.

The Moon is no exception; we can learn about the interior of our natural satellite from its deformation caused by the tidal force of Earth. The deformation had already been well known through several geodetic observations. However, models of the internal structure of the Moon as derived from past research could not account for the deformation precisely observed by the above lunar exploration programs. Therefore, the research team performed theoretical calculations to understand what type of internal structure of the Moon leads to the observed change of the lunar shape.

What the research team focused on is the structure deep inside the Moon. During the Apollo program, seismic observations were carried out on the Moon. One of the analysis results concerning the internal structure of the Moon based upon the seismic data indicates that the satellite is considered to consist mainly of two parts: the "core," the inner portion made up of metal; and the "mantle," the outer portion made up of rock. The research team has found that the observed tidal deformation of the Moon can be well explained if it is assumed that there is an extremely soft layer in the deepest part of the lunar mantle.

The previous studies indicated that there is the possibility that a part of the rock at the deepest layer inside the lunar mantle may be molten. This research result supports the above possibility, since partially molten rock becomes softer. The research has proven for the first time that the deepest part of the lunar mantle is soft, based upon the agreement between observation results and the theoretical calculations.

Furthermore, the research team also clarified that heat is efficiently generated by the tides in the soft part, deepest in the mantle. In general, a part of the energy stored inside a celestial body by tidal deformation is changed to heat. The heat generation depends on the softness of the interior. Interestingly, the heat generated in the layer is expected to be nearly at the maximum when the softness of the layer is comparable to that which the team estimated from the above comparison of the calculations and the observations.

This may not be a coincidence. Rather, the layer itself is considered to be maintained as the amount of the heat generated inside the soft layer is exquisitely well balanced with that of the heat escaping from the layer. Whereas previous research also suggests that some part of the energy inside the Moon due to the tidal deformation is changed to heat, the present research indicates that this type of energy conversion does not uniformly occur in the entire Moon, and occurs intensively only in the soft layer.

The research team believes that the soft layer is now warming the core of the Moon as the core seems to be wrapped by the layer, which is located in the deepest part of the mantle, and which efficiently generates heat. They also expect that a soft layer like this may efficiently have warmed the core in the past as well.

Concerning the future outlook for this research, Dr. Yuji Harada, the principal investigator of the research team, said, "I believe that our research results have brought about new questions. For example, how can the bottom of the lunar mantle maintain its softer state for a long time? To answer this question, we would like to further investigate the internal structure and heat-generating mechanism inside the Moon in detail. In addition, another question has come up: How has the conversion from the tidal energy to the heat energy in the soft layer affected the motion of the Moon relative to Earth, and also the cooling of the Moon? We would like to resolve those problems as well so that we can thoroughly understand how the Moon was born and has evolved."

Another investigator, Prof. Junichi Haruyama of Institute of Space and Aeronautical Science, Japan Aerospace Exploration Agency, mentioned the significance of this research, saying, "A smaller celestial body like the Moon cools faster than a larger one like Earth does. In fact, we had thought that volcanic activities on the Moon had already come to a halt. Therefore, the Moon had been believed to be cool and hard, even in its deeper parts.

However, this research tells us that the Moon has not yet cooled and hardened but is still warm. It even implies that we have to reconsider the question as follows: How have Earth and the Moon influenced each other since their births? That means this research not only shows us the actual state of the deep interior of the Moon but also gives us a clue for learning about the history of the system including both Earth and the Moon."

Shooting Marbles at 16,000 Miles per Hour

NASA scientist Bill Cooke is shooting marbles and he's playing "keepsies." The prize won't be another player's marbles, but knowledge that will help keep astronauts safe when America returns to the Moon in the next decade.

Cooke is firing quarter-inch-diameter clear shooters—Pyrex glass, to be exact—at soil rather than at other marbles. And he has to use a new shooter on each round because every 16,000 mph (7 km/s) shot destroys it (Fig. 10.11).

Fig. **10.11** A 30-cm-diameter crater plus spattered dust are all that's left after a test shot by the Ames Vertical Gun (Credit: NASA)

"We are simulating meteoroid impacts with the lunar surface," he explains. Cooke and others in the Space Environments Group at NASA's Marshall Space Flight Center have recorded the real thing many times. Their telescopes routinely detect explosions on the Moon when meteoroids slam into the lunar surface.

A typical flash involves "a meteoroid the size of a softball hitting the Moon at 27 km/s and exploding with as much energy as 70 kg of TNT."

"Mind you," says Cooke, "these are estimates based on a flash of light seen 400,000 km away. There's a lot of uncertainty in our calculations of speed, mass, and energy. We'd like to firm up these numbers."

That's where the marbles come in….

Cooke is using the Ames Vertical Gun Range at NASA's Ames Research Center in Mountain View, CA, to shoot marbles into simulated lunar soil. The firings allow him to calibrate what he sees on the Moon. His work is funded by NASA's Office of Safety and Mission Assurance.

"We measure the flash so we can figure out how much of the kinetic energy goes into light," he explained. "Once we know this luminous efficiency, as we call it, we can apply it to real meteoroids when they strike the Moon." High-speed cameras and a photometer (light meter) record the results.

The Ames Vertical Gun Range was built in the 1960s to support Project Apollo, America's first human missions to the lunar surface. The Ames gun can fire a variety of shapes and materials, even clusters of particles, at speeds from 0.5 to 7 km/s. The target chamber is usually pumped down to a vacuum and can be partially refilled to simulate atmospheres on other worlds or comets.

Equally important, the gun's barrel can be tilted to simulate impacts at seven different angles, from vertical to horizontal, since meteors rarely fly straight into the ground. Black powder propels the marble, and special valves capture the exhaust gases so they don't blow away the impact crater.

Cooke's experiments are being run in two rounds. The first set of 12 shots in October 2006 fired Pyrex glass balls into dust made from pumice, a volcanic rock, at up to 7 km/s. Follow-up experiments used JSC-1a lunar simulant, one of the "true fakes" developed from terrestrial ingredients to mimic the qualities of Moon soil.

Knowing the speed and mass of the projectile let Cooke scale the flash and estimate the energies of the softball-size meteoroids that hit the Moon at up to 72 km/s, more than six times the speed of the Ames gun. But luminous efficiency is just part of the question. A lot of the impact energy goes into shattering and melting the projectile—the main reason for using glass rather than metal—and then spraying debris everywhere.

"The ejecta kicked out from an impact can travel hundreds of miles," Cooke continued. "We need to know more about that if we are going to live on the lunar surface for months at a time." Because the Moon has virtually no atmosphere to slow down flying debris, particles land with the same speed that launched them from the impact site.

So you might dodge a bullet but still get caught by its shrapnel. And the question is: Are you more likely to get cut off at the ankles by debris spattered along the horizon, or hit from above by material on high, ballistic trajectories?

To gauge that danger, Cooke measured the speed and direction of secondary particles by the sheet-laser technique. Lenses and mirrors spread a laser beam into paper-thin sheets of light, so flying particles are briefly illuminated several times. The light traces then tell the size, direction, and speed of debris particles leaving an impact.

The technique requires a lot of image analysis, but it is cleaner and more accurate than the older way of hanging aluminum sheets in the chamber and counting holes.

The answers will help determine the kinds of shielding needed on exploration vehicles protecting humans where every day is for "keepsies."

Index

© Springer International Publishing Switzerland 2016

V.S. Foster, *Modern Mysteries of the Moon*, Astronomers' Universe,

DOI 10.1007/978-3-319-22120-5

FSC
www.fsc.org
MIX
Papier aus verantwortungsvollen Quellen
Paper from responsible sources
FSC® C105338